Cancer Etiology, Diagnosis and Treatments

Fanconi Anemia and Oxidative Stress

Mechanistic Background and Clinical Prospects

CANCER ETIOLOGY, DIAGNOSIS AND TREATMENTS

Additional books in this series can be found on Nova's website under the Series tab.

Additional e-books in this series can be found on Nova's website under the e-book tab.

CANCER ETIOLOGY, DIAGNOSIS AND TREATMENTS

FANCONI ANEMIA AND OXIDATIVE STRESS

MECHANISTIC BACKGROUND AND CLINICAL PROSPECTS

GIOVANNI PAGANO
EDITOR

New York

Library of Congress Cataloging-in-Publication Data

ISBN: 978-1-63482-297-8
Library of Congress Control Number: 2015932937

Published by Nova Science Publishers, Inc. † New York

Contents

Introduction

This book is to provide an up-to-date view of Fanconi Anemia (FA) as related to its dysmetabolic defects in terms of oxidative stress (OS) and mitochondrial dysfunction (MDF). A body of literature points to an involvement of OS in FA phenotype, as reported by molecular, *in vitro*, animal studies, and *in vivo*, from blood cells and biological fluids from FA patients.

The definition of FA is associated with DNA repair defect(s) relies on excess sensitivity of FA cells to interstrand crosslinking (ICL) agents, termed crosslinkers, as discussed in several chapters and, under a historical and comprehensive view, in Chapters 1 and 2. The excess sensitivity of FA cells to crosslinkers is discussed in Chapter 3 by referring to the literature pointing to anomalies of FA cells in crosslinker biotransformation and/or detoxification related to a number of redox mechanisms. Moreover, the body of literature associating redox-related dysfunctions of FA (FANC) proteins, is discussed in several chapters of this book. A specific focus on MDF as a major feature in FA phenotype is provided in Chapter 4, while Chapters 5 and 6 highlight the roles of OS-related aldehydes and of gluthathione in a number of diseases including FA. A particular focus on immunological anomalies, resulting in a proinflammatory condition in FA, is provided by Chapter 7.

A few introductory remarks about the roles of OS in the FA phenotype may be addressed as follows.

Oxidative Stress in FA Phenotype

An overall scenario for OS-related anomalies in FA cells, and in blood cells and biological fluids is summarized here (reviewed in Pagano et al. 2012;2013a).

The pioneering report by Ingrid Emerit (1977) found superoxide dismurase (SOD)-induced decrease of FA cell chromosomal instability, which was confirmed in mitocmycin C (MMC)-treated FA cells. Excess oxygen and iron sensitivity was reported in FA cells, both involving cell viability and G2 cell growth phase arrest (Joenje et al. 1981; Joenje and Oostra 1983; Nagasawa and Little 1983; Poot et al. 1996). Excess oxidative DNA damage was observed in FA lymphoblasts, which was enhanced in mitocmycin C (MMC)-exposed cells, and was partly attributed by Takeuchi and Morimoto (1993) to decreased catalase (CAT) activity, while thioredoxin (Trx) overexpression counteracted MMC- and diepoxybutane (DEB)-induced toxicity in FA fibroblasts (Ruppitsch et al. 1998).

Unconfined to *in vitro* testing, a series of independent studies conducted on blood cells and biological fluids from FA patients found an *in vivo* prooxidant state assessed on a set of redox endpoints, including oxidative DNA damage, GSH oxidation, WBC chemiluminescence, SOD and CAT expression, and excess aldehyde levels, including methylglyoxal and malondialdehyde (MDA) (Rumiantsev et al. 1989; Korkina et al. 1992; Degan et al. 1995; Pagano et al. 2004; Petrović et al. 2011).

A series of recent reports have focused on a deficiency of aldehyde dehydrogenase 2 (ALDH2) in FANCD2 cells and fancd2 (-/-) mice (Marietta et al. 2009; Langevin et al. 2011; Garaycoechea et al. 2012) relating ALDH2 deficiency to inability to counteract acetaldehyde toxicity. However, whether an ALDH2 defect does play a specific role in acetaldehyde detoxification awaits verification, in view of the excess *in vivo* levels of other aldehydes in FA patients (Pagano et al. 2004; Petrović et al. 2011), and due to low specificity of ALDH2 in aldehyde clearance, by e.g., counteracting 4-hydroxynonenal toxicity (Guo et al. 2013). One should remind the many possible factors leading to a prooxidant state, as in the case for oxidative DNA damage (8-OHdG), also detected in FA phenotype (both *in vitro* and *in vivo*) (Takeuchi and Morimoto 1993; Pagano et al. 2004; Petrović et al. 2011), displaying adverse effects including mutagenesis (Takeuchi et al. 1997; Halliwell 1999). Thus, an attempt to improve FA's clinical course by means of aldehyde detoxifying agents might not compensate the basic - and multi-faceted - pro-oxidant state featured in FA phenotype.

A few recent studies tested some different antioxidants in FA knock-out mice, reporting protective effects of some antioxidants (tempol and resveratrol, but not N-acetylcysteine) (Zhang et al. 2008; 2010; 2014). These findings might suggest an appropriate design of further animal studies and, possibly, of forthcoming clinical trials.

Mitochondrial Dysfunction

A highlight arises from the detection of mitochondrial dysfunction (MDF) in FA cell lines of four genetic subtypes (A, C, D2 and G), as reviewed in Chapter 4. This is closely related to redox impairment, pointing to multi-faceted links between MDF and OS in FA cellular phenotype (Mukhopadhyay et al. 2006; Kumari et al. 2014; Ravera et al. 2013). Alterations in mitochondrial morphology were reported by Kumari et al. (2014), including damaged mitochondrial membranes, abnormal shapes and possible events of mitophagy consistent with decreased mitochondrial transmembrane potential ($\Delta\Psi m$) and overproduction of ROS in FA cells. We have recently evaluated transcriptomal analyses on RNA obtained from bone marrow cells from FA patients *vs.* healthy volunteers, and identified significant changes in the expression of genes involved in bioenergetic pathways and in antioxidant activities (Pagano et al. 2013b).

This database on MDF is to date confined to FA cell lines, suggesting to plan studies focusing on any *in vivo* MDF endpoints in FA patients.

Redox-related Roles of FANC Proteins

Some FANC proteins have been associated with several redox-related activities and/or interactors.

Independent studies by Kruyt et al. and by Futaki et al. found that FANCC and FANCG proteins bind to and regulate the activity of cytochrome-P450 (Kruyt et al. 1998; Futaki et al. 2002) microsomal membrane proteins involved in electron transfer and long recognized as effectors in MMC bioactivation (Keyes et al. 1984).

A direct involvement of BRCA1 and BRCA2 (FANCD1) proteins was reported in the repair of 8-oxoguanine damage, thus implying their direct role in redox pathways (Le Page et al. 2000), as oxidation of guanine is a well-known parameter of OS and key factor in cancer proneness (Halliwell 1999). Other investigations found functional interactions of FANCC and FANCG proteins with NADPH cytochrome P450 reductase, and by redox regulation of GSTP1 (Cumming et al. 2001). Park et al. (2004) found that FANCA and FANCG proteins were sensitive to redox conditions, since their monomers became multimers following H_2O_2 treatment. A series of studies by Qishen Pang's group reported on the manifold involvement of the FA pathway in counteracting OS through selective protection of antioxidant defense (Zhang et al. 2005; 2007; Du et al. 2012).

An extensive database points to the multiple links of the FANCJ (FANCJ/BACH1/BRIP1) protein with OS. This protein displays functional interaction with BRCA1 in breast and ovary cancer suppression (Levran et al. 2005; Cantor and Andreassen 2006) and, in turn, BRCA1 is known to be involved in OS- related mechanisms (Le Page et al. 2000; Vurusaner et al. 2012). Moreover, FANCJ/BACH1/BRIP1 competes with Nrf2 leading to negative regulation of the antioxidant response element (ARE)-mediated NAD(P)H: quinone oxidoreductase 1 gene expression and induction in response to antioxidants (Dhakshinamoorthy et al. 2005). FANCJ/BACH1/BRIP1 features a helicase activity and an ATPase activity, a two-fold activity that is recognized for other RECQ helicases that are associated with ATPase activity (Monnat 2010). FANCJ/BACH1/BRIP1 regulates heme oxygenase-1 (Kitamuro et al. 2003; Okada et al. 2010), senses oxidative DNA damage and is stimulated by replication protein A to unwind the damaged DNA substrate in a strand-specific manner (Suhasini et al. 2009). Altogether, the database on interactions of several FANC proteins points to multiple molecular interactions of at least five FANC (A, C, D2, G and J) proteins with redox-related mechanisms.

Conclusion

This book provides an overall view of the state-of-art on the links of FA phenotype with OS and MDF, while questioning the validity of the universally accepted theory of DNA repair defects (Chapter 3). Especially, this book may provide the readers with working hypotheses in view of elucidating the *in vivo* relevance of OS and MDF in FA patients, with the ultimate goal of undertaking evidence-based clinical trials enabling the clinicians to improve FA's clinical course.

References

Cantor, S.B., & Andreassen, P.R. (2006). Assessing the link between BACH1 and BRCA1 in the FA pathway. *Cell Cycle 5*, 164-167.

Cumming, R.C., Lightfoot, J., Beard, K., Youssoufian, H., O'Brien, P.J., & Buchwald, M. (2001). Fanconi anemia group C protein prevents apoptosis in hematopoietic cells through redox regulation of GSTP1. *Nature Med. 7*, 814-820.

Degan, P., Bonassi, S., De Caterina, M., Korkina, L.G., Pinto, L., Scopacasa, F., Zatterale, A., Calzone, R., & Pagano, G. (1995) In vivo accumulation of 8-hydroxy-2'-deoxyguanosine in DNA correlates with release of reactive oxygen species in Fanconi's anaemia families. *Carcinogenesis 16*, 735-742.

Dhakshinamoorthy, S., Jain, A.K., Bloom, D.A., & Jaiswal, A.K. (2005) Bach1 competes with Nrf2 leading to negative regulation of the antioxidant response element (ARE)-mediated NAD(P)H:quinone oxidoreductase 1 gene expression and induction in response to antioxidants. *J. Biol. Chem. 280*, 16891-16900.

Du, W., Rani, R., Sipple, J., Schick, J., Myers, K.C., Mehta, P., Andreassen, P.R., Davies, S.M., & Pang, Q. (2012). The FA pathway counteracts oxidative stress through selective protection of antioxidant defense gene promoters. *Blood 119*, 4142-4151.

Futaki, M., Igarashi, T., Watanabe, S., Kajigaya, S., Tatsuguchi, A., Wang, J., & Liu, J.M. (2002). The FANCG Fanconi anemia protein interacts with CYP2E1: possible role in protection against oxidative DNA damage. *Carcinogenesis 23*, 67-72.

Garaycoechea, J.I., Crossan, G.P., Langevin, F., Daly, M., Arends, M.J., & Patel, K.J. (2012). Genotoxic consequences of endogenous aldehydes on mouse haematopoietic stem cell function. *Nature 489*, 571-575.

Guo, J.M., Liu, A.J., Zang, P., Dong, W.Z., Ying, L., Wang, W., Xu, P., Song, X.R., Cai, J., Zhang, S.Q., Duan, J.L., Mehta, J.L., & Su, D.F. (2013). ALDH2 protects against stroke by clearing 4-HNE. *Cell Res. 23*, 915-930.

Halliwell, B. (1999). Oxygen and nitrogen are pro-carcinogens. Damage to DNA by reactive oxygen, chlorine and nitrogen species: measurement, mechanism and the effects of nutrition. *Mutat. Res. 443*, 37-52.

Joenje, H., Arwert, F., Eriksson, A.W., de Koning, H., & Oostra, A.B. (1981). Oxygen-dependence of chromosomal aberrations in Fanconi's anaemia. *Nature 290*, 142-143.

Joenje, H., & Oostra, A.B. (1983). Effect of oxygen tension on chromosomal aberrations in Fanconi anaemia. *Hum. Genet. 65*, 99-101.

Keyes, S.R., Fracasso, P.M., Heimbrook, D.C., Rockwell, S., Sligar, S.G., & Sartorelli, A.C. (1984). Role of NADPH:cytochrome c reductase and DT-diaphorase in the biotransformation of mitomycin C1. *Cancer Res. 44*, 5638-5643.

Kitamuro, T., Takahashi, K., Ogawa, K., Udono-Fujimori, R., Takeda, K., Furuyama, K., Nakayama, M., Sun, J., Fujita, H., Hida, W., Hattori, T., Shirato, K., Igarashi, K., & Shibahara, S. (2003). Bach1 functions as a hypoxia-inducible repressor for the heme oxygenase-1 gene in human cells., 9125-9133.

Korkina, L.G., Samochatova, E.V., Maschan, A.A., Suslova, T.B., Cheremisina, Z.P., & Afanas'ev, I.B. (1992). Release of active oxygen radicals by leukocytes of Fanconi Anemia patients. *J. Leukoc. Biol. 52*, 357-362.

Kruyt, F.A., Hoshino, T., Liu, J.M., Joseph, P., Jaiswal, A.K., & Youssoufian, H. (1998). Abnormal microsomal detoxification implicated in Fanconi anemia group C by interaction of the FAC protein with NADPH cytochrome P450 reductase. *Blood 92*, 3050-3056.

Kumari, U., Jun, W.Y., Bay, B.H., & Lyakhovich, A. (2014). Evidence of mitochondrial dysfunction and impaired ROS detoxifying machinery in Fanconi anemia cells. *Oncogene 33*, 165-172.

Langevin, F., Crossan, G.P., Rosado, I.V., Arends, M.J., & Patel, K.J. (2011). Fancd2 counteracts the toxic effects of naturally produced aldehydes in mice. *Nature 475*, 53-58.

Le Page, F., Randrianarison, V., Marot, D., Cabannes, J., Perricaudet, M., Feunteun, J., & Sarasin, A. (2000). BRCA1 and BRCA2 are necessary for the transcription-coupled repair of the oxidative 8-oxoguanine lesion in human cells. *Cancer Res. 60*, 5548-5552.

Levran, O., Attwooll, C., Henry, R.T., Milton, K.L., Neveling, K., Rio, P., Batish, S.D., Kalb, R., Velleuer, E., Barral, S., Ott, J., Petrini, J., Schindler, D., Hanenberg, H., & Auerbach, A.D. (2005). The BRCA1-interacting helicase BRIP1 is deficient in Fanconi anemia. *Nat. Genet. 37*, 931-933.

Marietta, C., Thompson, L.H., Lamerdin, J.E., & Brooks, P.J. (2009) Acetaldehyde stimulates FANCD2 monoubiquitination, H2AX phosphorylation, and BRCA1 phosphorylation in human cells in vitro: implications for alcohol-related carcinogenesis. *Mutat. Res. 664*, 77-83.

Monnat, R.J. Jr. (2010). Human RECQ helicases: roles in DNA metabolism, mutagenesis and cancer biology. *Semin. Cancer Biol. 20*, 329-339.

Mukhopadhyay, S.S., Leung, K.S., Hicks, M.J., Hastings, P.J., Youssoufian, H., & Plon, S.E. (2006). Defective mitochondrial peroxiredoxin-3 results in sensitivity to oxidative stress in Fanconi anemia. *J. Cell Biol. 175*, 225-235.

Nagasawa, H., & Little, J.B. (1983). Suppression of cytotoxic effect of mitomycin-C by superoxide dismutase in Fanconi's anemia and dyskeratosis congenita fibroblasts. *Carcinogenesis 4*, 795-799.

Nordenson, I. (1977). Effect of superoxide dismutase and catalase on spontaneously occurring chromosome breaks in patients with Fanconi's anemia. *Hereditas 86*, 147-150.

Okada, S., Muto, A., Ogawa, E., Nakanome, A., Katoh, Y., Ikawa, S., Aiba, S., Igarashi, K., & Okuyama, R. (2010) Bach1-dependent and -independent regulation of heme oxygenase-1 in keratinocytes. *J. Biol. Chem. 285*, 23581-23589.

Pagano, G., Degan, P., d'Ischia, M., Kelly, F.J., Pallardó, F.V., Zatterale, A., Anak, S.S., Akisik, E.E., Beneduce, G., Calzone, R., De Nicola, E., Dunster, C., Lloret, A., Manini, P., Nobili, B., Saviano, A., Vuttariello, E., & Warnau, M. (2004). Gender- and age-related distinctions for the in vivo prooxidant state in Fanconi anaemia patients. *Carcinogenesis 25*, 1899-1909.

Pagano, G., Talamanca, A.A., Castello, G., Pallardó, F.V., Zatterale, A., & Degan, P. (2012). Oxidative stress in Fanconi anaemia: from cells and molecules toward prospects in clinical management. *Biol. Chem. 393*, 11-21.

Pagano, G., Aiello Talamanca, A., Castello, G., d'Ischia, M., Pallardó, F.V., Petrović, S., Porto, B., Tiano, L., & Zatterale, A. (2013a). From clinical description to in vitro and animal studies, and backwards to patients: Oxidative stress and mitochondrial dysfunction in Fanconi anemia. *Free Radic. Biol. Med. 58*, 118-125.

Pagano G., Talamanca A.A., Castello G., d'Ischia M., Pallardó F.V., Petrović S., Porto B., Tiano L., & Zatterale A. (2013b). Bone marrow cell transcripts from Fanconi anaemia patients reveal in vivo alterations in mitochondrial, redox and DNA repair pathways. *Eur. J. Haematol. 91*, 141-151.

Park, S.J., Ciccone, S.L., Beck, B.D., Hwang, B., Freie, B., Clapp, D.W., & Lee, S.H. (2004). Oxidative stress/damage induces multimerization and interaction of Fanconi anemia proteins. *J. Biol. Chem. 279*, 30053-30059.

Petrović, S., Leskovac, A., Kotur-Stevuljević, J., Joksić, J., Guc-Scekić, M., Vujić, D., & Joksić, G. (2011). Gender-related differences in the oxidant state of cells in Fanconi anemia heterozygotes. *Biol. Chem. 392*, 625-632.

Poot, M., Gross, O., Epe, B., Pflaum, M., & Hoehn, H. (1996). Cell cycle defect in connection with oxygen and iron sensitivity in Fanconi anemia lymphoblastoid cells. *Exp. Cell Res. 222*, 262-268.

Ravera, S., Vaccaro, D., Cuccarolo, P., Columbaro, M., Capanni, C., Bartolucci, M., Panfoli, I., Morelli, A., Dufour, C., Cappelli, E., & Degan P. (2013). Mitochondrial respiratory chain Complex I defects in Fanconi anemia complementation group A. *Biochimie 95*, 1828-1837.

Rumiantsev, A.G., Samochatova, E.V., Afanas'ev, I.B., Korkina, L.G., & Suslova, T.B. (1989). The role of free oxygen radicals in the pathogenesis of Fanconi's anemia. *Ter. Arkh. 61*, 32-36.

Ruppitsch, W., Meisslitzer, C., Hirsch-Kauffmann, M., & Schweiger, M. (1998). Overexpression of thioredoxin in Fanconi anemia fibroblasts prevents the cytotoxic and DNA damaging effect of mitomycin C and diepoxybutane. *FEBS Lett. 422*, 99-102.

Suhasini, A.N., Sommers, J.A., Mason, A.C., & Voloshin, O.N., Camerini-Otero, R.D., Wold, M.S., & Brosh, R.M. Jr. (2009). FANCJ helicase uniquely senses oxidative base damage in either strand of duplex DNA and is stimulated by replication protein A to unwind the damaged DNA substrate in a strand-specific manner. *J. Biol. Chem. 284*, 18458-18470.

Takeuchi, T., & Morimoto, K. (1993). Increased formation of 8-hydroxydeoxyguanosine, an oxidative DNA damage, in lymphoblasts from Fanconi's anemia patients due to possible catalase deficiency. *Carcinogenesis 14*, 1115-1120.

Takeuchi, T., Matsugo, S., & Morimoto, K. (1997) Mutagenicity of oxidative DNA damage in Chinese hamster V79 cells. *Carcinogenesis 18*, 2051-2055.

Vurusaner, B., Poli, G., & Basaga, H. (2012). Tumor suppressor genes and ROS: complex networks of interactions. *Free Radic. Biol. Med. 52*, 7-18.

Zhang, Q.S., Eaton, L., Snyder, E.R., Houghtaling, S., Mitchell, J.B., Finegold, M., Van Waes, C., & Grompe, M. (2008). Tempol protects against oxidative damage and delays epithelial tumor onset in Fanconi anemia mice. *Cancer Res. 68*, 1601-1608.

Zhang, Q.S., Marquez-Loza, L., Eaton, L., Duncan, A.W., Goldman, D.C., Anur, P., Watanabe-Smith, K., Rathbun, R.K., Fleming, W.H., Bagby, G.C., & Grompe, M. (2010). Fancd2-/- mice have hematopoietic defects that can be partially corrected by resveratrol. *Blood 116*, 5140-5148.

Zhang, Q.S., Marquez-Loza, L., Sheehan, A.M., Watanabe-Smith, K., Eaton, L., Benedetti, E., Major, A., Schubert, K., Deater, M., Joseph, E., & Grompe, M. (2014). Evaluation of resveratrol and N-acetylcysteine for cancer chemoprevention in a Fanconi anemia murine model. *Pediatr. Blood Cancer 61*, 740-742.

Zhang, X., Li, J., Sejas, D.P., & Pang, Q. (2005). Hypoxia-reoxygenation induces premature senescence in FA bone marrow hematopoietic cells. *Blood 106*, 75-85.

Zhang, X., Sejas, D.P., Qiu, Y., Williams, D.A., & Pang, Q. (2007). Inflammatory ROS promote and cooperate with the Fanconi anemia mutation for hematopoietic senescence. *J. Cell Sci. 120(Pt 9)*, 1572-83.

Giovanni Pagano
Italian National Cancer Institute, CROM
I-83013 Mercogliano (AV), Italy; E-mail: gbpagano@tin.it

In: Fanconi Anemia and Oxidative Stress
Editor: Giovanni Pagano
ISBN: 978-1-63482-297-8

Chapter 1

Fanconi Anemia Clinical and Genetic Heterogeneity

Adriana Zatterale*

Head Emeritus, Genetics Department, ASL Napoli 1, Naples, Italy

Abstract

Fanconi Anemia (FA) is a rare recessive disease, first reported in 1927 by Guido Fanconi as familial, infantile anemia, now defined as a chromosomal instability (CI) syndrome, showing a wide clinical and genetic heterogeneity. Genetically FA can be caused by mutations in at least 16 different genes, mostly cooperating in a pathway whose functions are not yet fully known and somehow interacting with proteins produced by genes which, when mutated, cause other CI syndromes, as Ataxia Teleangiectasia, Bloom syndrome and Nijmegen Breakage Syndrome. This pathway is likely to relate to DNA damage repair and to oxidative stress. Clinically FA patients present bone marrow failure at variable time in life, typically beginning in infancy as platelet deficiency and progressive to pancytopenia and death. The patients can show variable congenital malformations and proneness to hematologic and solid malignancies. The FA cellular phenotype is characterized by a G2 cell cycle phase delay and by CI, especially present as rearrangement figures (triradial and quadriradial figures made by non-homologous chromosomes). The CI is both spontaneous and/or induced by alkylating DNA cross-linking agents, such as mitomycin C (MMC) or the more specific diepoxybutane (DEB).

The cytogenetic DEB-test makes the FA diagnosis possible in patients not showing typical malformations or still asymptomatic and also in patients not showing spontaneous CI at standard cytogenetic evaluation. At present time the hematopoietic stem cell transplantation represents the only effective treatment for FA, but unfortunately it can neither improve the patient's growth rate nor the predisposition to neoplasia.

Keywords: chromosomal instability, FANC proteins, mitomycin C, diepoxybutane, bone marrow failure, genetic heterogeneity

* Corresponding author: E-mail: azatt@tin.it.

Introduction

In 1927 a young Swiss pediatrician, working at the Children's Hospital in Zurich, reported on a kindred of three brothers affected by congenital anomalies and a lethal anemia whose onset was between 5 and 7 years of age (Fanconi 1927). Few years later, this syndromic condition was named after him as *Fanconi anemia* (FA). A similar hematologic condition without congenital malformations was reported and regarded as a *Fanconi anemia variant* (Estren and Dameshek 1947; Zaizov et al. 1969; Nowell et al. 1984). The subsequent knowledge of the wide phenotypic variability of the syndrome made this differentiation obsolete.

The present classification sets this rare genetic disease among the *Chromosomal Instability Syndromes*. FA is characterized by progressive bone marrow failure, frequent physical and developmental anomalies and proneness to leukemia and solid tumors, as well as CI.

Clinical Aspects

Congenital and Developmental Abnormalities

A newborn with FA may - or may not - present some congenital malformations, affecting any of the organ systems. Strongly suggestive of FA (but not exclusive) are mono- or bilateral defects of the radial ray, as supernumerary or bifid thumbs, thumb and/or radius aplasia/hypoplasia (Figure 1). Different skeletal anomalies, as vertebral abnormalities, gastrointestinal malformations, as tracheo-esophageal fistula and anal atresia, cardiac and renal abnormalities may lead to misdiagnose these children as affected by different genetic diseases (e.g., TAR syndrome, Holt-Oram syndrome, Seckel syndrome, very often VACTER or VACTERL associations, Roberts syndrome, Warsaw breakage syndrome) (van der Lelij et al. 2010).

A low birth weight owing to prenatal growth retardation is almost the rule.

Other malformations may affect the ear (morphologic anomalies, hearing loss), the eye (microphthalmia, strabismus, cataracts), and/or the central nervous system (from hydrocephalus to microcephaly), and may show up in course of time.

Skin color anomalies, either diffuse hyperpigmentation (brownish skin, suntanned appearance) or café-au-lait spots, as well as localized hypopigmentation (Figure 2), are frequent features.

The stature can differ in the same affected kindred but in at least half of patients it stays on a short level, even below the third centile. The short stature is often due to endocrine hypofunction (growth hormone or insulin deficiencies, hypothyroidism) (Wajnrajch et al. 2001; Giri et al. 2007).

Developmental abnormalities may be present, varying from failure to thrive to delayed puberty, with hypogonadism in males, early menopause in females, both sexes showing reduced fertility (abnormal spermatogenesis in males; rare and complicated pregnancies in females).

The malformation pattern is highly variable among families (inter-familial variability), as well as among affected siblings (intra-familial variability), showing differences even between monozygotic twins, then ultrasound examination cannot be used as prenatal diagnostic tool in FA families.

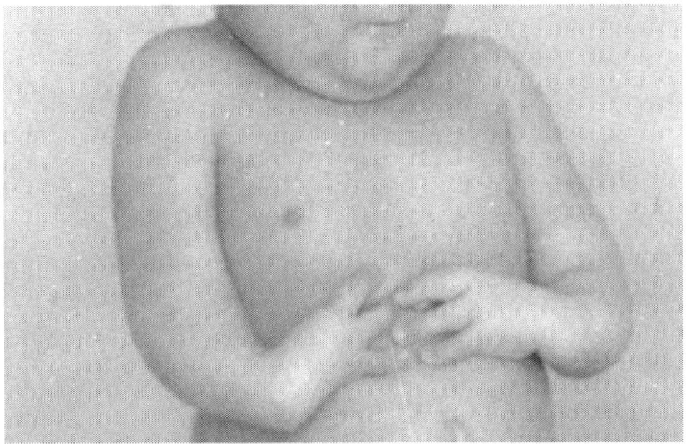

Figure 1. Radius and thumb aplasia, widely spaced nipples and micrognathia in a child with FA.

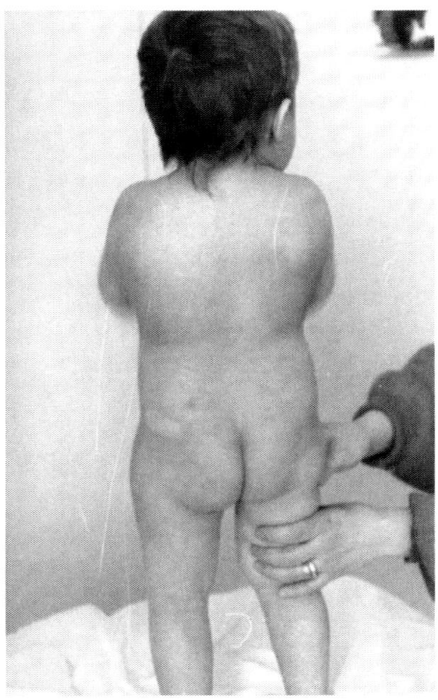

Figure 2. Localized hypo- and hyper-pigmentation in a FA child.

However, 25-40% of patients do not have any major malformations; therefore FA cannot be suspected in them until the onset of hematologic symptoms (Alter 1993).

Hematologic Aspects

Very rare is the case of a FA newborn showing hematologic problems, that instead usually appear during the infancy (median age: 7 years; range: birth to 40's) with recurrent bruising, petechiae, bleeding and fatigue, due to an incoming thrombocytopenia (platelets less than 100×10^9/L) or recurring respiratory tract infections, due to a leukopenia, mostly neutropenia (neutrophils less than 1×10^9/L).

Increased fetal hemoglobin with no hemoglobinopathy and erythrocyte macrocytosis can also be found at the peripheral blood examination.

Bone marrow (BM) becomes hypoplastic in a variable time, progressing to aplastic anemia and pancytopenia (affecting at least 2-3 blood cell lineages) (Butturini et al. 1994).

Clonal chromosomal aberrations may appear – in few months to many years – during BM failure, often involving chromosomes 1, 3, 5, 6, 7, 8, 11, 21, a sign of a subsequent evolution – after some weeks to few years – to a myelodysplastic syndrome (MDS) and/or a leukemia, typically an acute myelogenous leukemia (AML) (Quentin et al. 2011). The actuarial risk of MSD or AML evolution is about 40-60% by 40 years of age, being higher in patients with a history of clonal chromosomal aberrations (Auerbach and Allen 1991).

Cytogenetic Findings

In 1964 the evidence of a peculiar cytogenetic finding on FA cells was reported (Schroeder et al. 1964; Bloom et al. 1966), defined as CI and characterized by significant increase of chromatid and/or chromosome breakages and rearrangements, some of which being very typical (triradial and quadriradial figures, mostly asymmetrical, that means formed by at least two non-homologous chromosomes) in a cytogenetic standard culture obtained from phytohemagglutinin-stimulated peripheral blood lymphocytes (Figures 3-4).

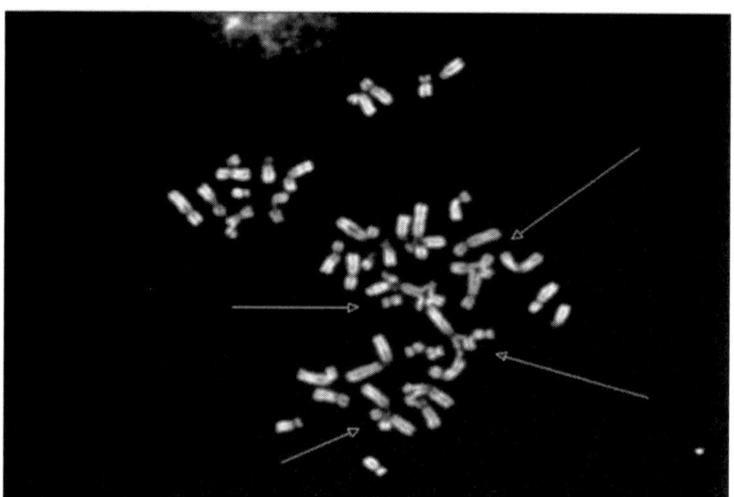

Figure 3. Metaphasic chromosomes forming asimmetrical quadriradial figures and complex rearrangements in FA peripheral blood lymphocytes.

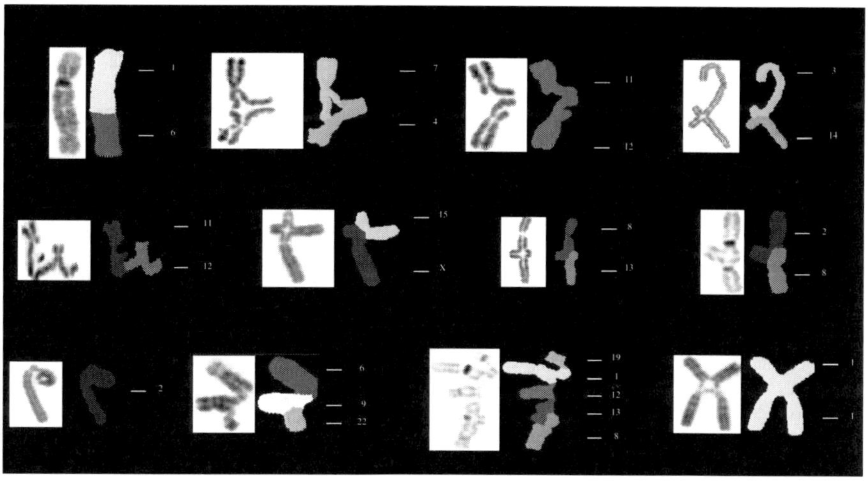

Figure 4. Spectral Karyotyping (SKY) in FA, showing that rearrangements are prevalently made by two or more non-homologous chromosomes (asimmetrical figures). The bottom on the right is a simmetrical quadriradial figure.

Unfortunately some patients, even though affected by FA typical birth defects and BM failure, lacked in showing spontaneous CI *in vitro*, then treatment in culture with specific agents had been tried to enhance or induce CI of FA cells.

The first successful attempt brought into use MMC, a DNA bifunctional crosslinking, alkylating chemical (Sasaki and Tonomura 1973). After few years DEB was tested successfully (Auerbach and Wolman 1976; Auerbach 1993), becoming the most specific agent for FA laboratory diagnosis.

FA patients developing MDS or AML seem to show specific cytogenetic aberrations and recent evidence suggests a role of oxidative stress damage on FA hematopoietic cell dysfunction (Du et al. 2008).

Cancer Proneness

Cancer predisposition in genetic diseases may be originated through many mechanisms (loss of a gene function creating alterations in metabolic pathways normally suppressing oncogenesis, cell cycle regulation and checkpoint imbalance, immune deficiencies, etc.) (Zatterale et al. 1999). Whatever should be the basic DNA defect in a FA patient, the DNA persisting damage easily leads to neoplastic evolution.

Indeeed FA patients show a high predisposition both to hematologic malignancies, mainly MDS and AML, and to solid tumors, generally developed at an unusual early age without any evidence of an environmental predisposing agent exposure (Rosenberg et al. 2003).

The highest frequency is of squamous cell carcinomas of head and neck (500- to 700-fold higher risk than general population) and anogenital mucosae, which have been associated with the inactivation of p53 by HPV oncoproteins (van Zeeburg et al. 2004). Therefore virus-related carcinogenesis, a multifactorial mechanism, should be taken into account.

Cancer therapy in FA patients is complicated by their hypersensitivity to radiotherapy and to some chemotherapic agents.

Genetic Aspects

Inheritance

Since the first report by Guido Fanconi it became evident that the disease should be genetic, with a recessive mode of inheritance, because affecting some sibs with normal parents being healthy carriers. The genetic heterogeneity, supposed on the basis of the high clinical variability, has been confirmed by complementation studies on hybrid cell lines, leading in time to prove the existence at present of at least 16 genes located on different autosome arms except one (FANCB) located on the short arm of the X chromosome. In this case females are the FA carriers, transmitting the illness to their sons only.

FANC Genes

The characteristics of the already known FA genes, whose mutations have been proven to cause a FA phenotype, are summarized in Table 1.

Surprisingly, FANCD1 has been found identical to one of the genes predisposing to hereditary breast and ovarian cancer, BRCA2 (D'Andrea 2010). FANCJ is the same as BRIP1, a BRCA1 interacting helicase (Levitus et al. 2005; Litman et al. 2005); FANCN is the same as PALB2, a partner and localizer of BRCA2 (Reid et al. 2007); FANCO corresponds to a homologous repair factor, already known as RAD51C (Meindl et al. 2010; Vaz et al. 2010).

Table 1. FANC genes

Subtype	Gene	Locus	Prevalence in FA Patiens	Protein (kDa)
A	FANCA	16q24.3	≈66%	163
B	FANCB	Xp22.2	≈2%	95
C	FANCC	9q22.32	≈10%	63
D1	FANCD1/BRCA2	13q13.1	≈2%	380
D2	FANCD2	3p25.3	≈3%	155-162
E	FANCE	6p21.31	≈2%	60
F	FANCF	11p14.3	≈2%	42
G	FANCG/ XRCC9	9p13.3	≈9%	68
I	FANCI (KIAA1794)	15q26.1	<2%	142-150
J	FANCJ/BACH1/BRIP1	17q23.2	<2%	150
L	FANCL/PHF9/POG	2p16.1	Rare	43
M	FANCM/Hef	14q21.2	Rare	250
N	FANCN/PALB2	16p12.2	Rare	130
O	FANCO/RAD51C	17q22	Rare	43
P	FANCP/SLX4	16p13.3	Rare	200
Q	FANCQ/ERCC4/XPF	16p13.12	Rare	104

All these genes – and probably many others – cooperate each one expressing a pleiotropic effect on phenotype, too complex to explain to date.

Roles of FANC Proteins

The products encoded by FANC genes have been demonstrated to work in a common cellular pathway, and their functions are not yet fully explored.

Some of them (FANCA, FANCB, FANCC, FANCE, FANCF, FANCG, FANCL, FANCM) are assembled in a nuclear complex (FA core complex) which mediates the monoubiquitylation of FANCD2 and FANCI (Sims et al. 2007), that once monoubiquitylated constitute a heterodimer (ID complex) (Joo et al. 2011) thereafter interacting with downstream FANC proteins (FANCD1/BRCA2, FANCJ/BACH1/BRIP1, FANCN/PALPB2, FANCO/ RAD51C, FANCP/SLX4) (Hussain et al. 2004; Wang et al. 2004) and some associated proteins as the other gene predisposing to hereditary breast and ovarian cancer, BRCA1. The monoubiquitylated FANCD2 acts also as a signal for the recruitment of downstream effector proteins as FAN1 (Fanconi associated nuclease 1) (Liu et al. 2010).

Therefore pathologic mutations in one of the eight genes (FANCA, FANCB, FANCC, FANCE, FANCF, FANCG, FANCL, FANCM) whose products work upstream the pathway, prevent the monoubiquitylation of FANCD2 and FANCI, hence arrest the pathway functioning.

Many so-called FA-associated proteins are integral part of the FA pathway functioning but are not considered FANC proteins until a pathologic mutation is found in a FA patient.

The exact role of the FA pathway is far from being completely known. Most FA core complex proteins do not show any enzymatic domain, therefore they may function as a scaffold. Evidence exists for a role in DNA repair, mostly of downstream FANC proteins on homologous recombination contributing to repair the interstrand cross-links which cause replication fork stalling. This role is also proven by the established interactions of FA pathway proteins with some proteins produced by genes which, when mutated, cause other CI syndromes, as Ataxia Teleangiectasia (AT), Bloom syndrome (BS), and Nijmegen Breakage Syndrome (NBS).

A body of evidence exists however on the hypersensitivity of FA cells to oxidative stress (OS) (Pagano et al. 2013). Interestingly the DNA repair is exerted during the S phase of the cellular division, but the FA cells show a delay during G2 phase. Therefore it may be speculated if in FA cells a DNA repair defect during S phase makes FA cells hypersensitive to oxidative damage, or rather their hypersensitivity to reactive oxygen species (ROS) causes an unbearable redox imbalance, i.e., a damage that they cannot cope with. Actually the metabolic pathways normally generate a number of DNA damaging molecules, as the ROS, which constitute the endogenous burden to be added to the exogenous DNA damage coming from the environmental OS.

Several early and recent studies provided insight regarding this topic. Since 1977 some authors reported on a significant CI decrease in cultured FA cells treated with superoxide dismutase, an enzyme which defends cells against oxygen toxicity (Nordenson 1977) and CI dependent on oxygen level (Joenje et al. 1981; Joenje and Oostra 1983). At present, even though most of research has been addressed to investigate the role of FA pathway in DNA damage repair, we know that some FANC proteins have been found to be associated with redox-related activities (FANCA, FANCC and FANCG interact with cytochrome P450-related activities and/or respond to oxidative damage, FANCD2 responds to OS interacting with forkhead box O3, FANCG - also found in mitochondria – interacts with PRDX3, FANCJ is a repressor of heme-oxigenase-1) as well as with oxidative metabolism, transcription regulation, signaling, cell cycle checkpoint control, intracellular transport and apoptosis programming (Pagano et al. 2013). It remains to be verified whether redox imbalance, as all the different functions which FA pathway is involved in, is likely to be the consequence or rather the cause of the genetic defect.

Particularly it is still unknown which role the FANC genes play in regulating the growth, differentiation and survival of normal hematopoietic stem cells. Therefore we still ignore how the pancytopenia, the clinical FA hallmark, is originated lacking a FA pathway product.

Genotype-Phenotype Correlations

Genotype-phenotype correlations have been attempted, not a simple task given the extreme clinical variability and the high genetic heterogeneity, complicated by the wide number of mutation types found in patients, some of them leading to differently partially functioning genes. Moreover, it has to be taken into account that, as the FA pathway works in a genetic network, gene-genes, gene-environment and epigenetic effects should act as modifiers of the mutation impact.

Correlations have been approached for the three more common complementation groups, based on International and National registry reports, the group A (Faivre et al. 2000), the group G (Demuth et al. 2000), and the group C (Kutler et al. 2003). Tentatively, we may consider that FANCD1 and FANCN patients show earlier onset and higher risk of malignancies, whereas FANCA patients usually show milder clinical manifestations than FANCC and FANCG patients. This kind of evidence can be useful for clinicians for prognostic purposes, to reserve more precocious and hazardous interventions - as hematopoietic stem cell transplantation (HSCT) from an unrelated donor – and an appropriate cancer surveillance to the patients at higher risk (Barrett et al. 1977; Broxmeier et al. 1989).

Mosaicism

A genetic mosaicism is defined as the presence in the same organism of at least two somatic cell lines genetically different. In FA this situation can be revealed by the presence in culture of two or more populations of lymphocytes characterized by a different degree of sensitivity to the alkylating agents used to test the CI (Lo Ten Foe et al. 1997). The FA immortalized lymphoblastoid cell lines often show a "reverted" phenotype with aging, i.e., they lose the sensitivity that the lymphocytes of the same patient are still keeping, because of a spontaneous mutation occurred in a cell line subpopulation, reverting the pathologic mutation to the wild-type DNA sequence.

The DNA instability may lead in rare cases also *in vivo* to a restoration of one FA gene copy which was mutated, the reverted cells contributing to relatively improve the hematologic phenotype and the disease course because of the clonal expansion of the reverted cells, due their survival advantage promoted by selection pressure on cells correctly replicating.

The mosaicism can be originated by a spontaneous point mutation or an intragenic mitotic recombination at a FA locus in a compound heterozygous patient. Other mechanisms, as modifier genes involvement, could be postulated (Davies et al. 2005). The mosaicism is a phenomenon observed also in other CI syndromes (e.g.: BS, AT). It is always a stochastic event, as proved by the existence of discordant monozygotic FA twins, one mosaic, the other not.

Diagnostic Tools

A correct and precocious diagnosis, that plays a very important role in fighting FA, requires some consecutive steps, first of all the CI test, namely the DEB-test.

DEB-Test

FA diagnosis, due to highly variable clinical presentation of the disease, needs a specific laboratory test, based on the increased CI that cultured FA cells show in response to alkylating agents (clastogenic effect), whereas their sensitivity to radiation is close to normal, as well as the sister chromatid exchanges (SCE).

The possibility of a hematopoietic somatic mosaicism has to be considered, as no correlation is demonstrated so far between phenotype severity and CI degree.

The DEB-test is normally performed on peripheral blood lymphocytes. It can be performed also on skin fibroblasts, if a mosaicism is suspected, on bone marrow fibroblasts, and, for prenatal diagnosis purposes, on fetal cells (Auerbach et al. 1979) obtained usually by amniocentesis or by chorionic villi sampling or, at later pregnacy stages, by percutaneous umbilical cord blood sampling. Some accredited laboratories prefer to use MMC, though less specific, because DEB is a highly volatile carcinogen, moreover it is hygroscopic losing activity upon contact with water or plastic devices, then requires great caution to be handled.

The CI diagnosis requires a specific expertise beyond the basic cytogenetic experience, and the use of specific controls, consisting in FA cell lines not showing spontaneous CI but only induced by clastogenic agents.

Cell Cycle Studies

FA lymphocytes exposed to cross-linking agents show a significant elongation of the G2 phase during the mitotic division (Dutrillaux et al. 1982), due to an increased activity of a checkpoint kinase (CHK1). This abnormality of the cell cycle kinetics can be demonstrated by flow cytometric techniques requiring a specialized instrumentation and experience.

Complementation and Molecular Studies

To subtype FA patients, their cells are transduced with retroviruses carrying cDNAs, each one of them complementing a different FA mutated gene, i.e., correcting the CI due to the hypersensitivity to alkylating agents or restoring the monoubiquitylation if the mutated gene is part of the upstream complex.

The difficulties in performing complementation tests may be surpassed in some cases using a proteomic approach based on the possibility to distinguish the ubiquitylated from the non-ubiquitylated FANCD2 forms by Western Blotting (WB). This relatively simple and inexpensive technique leads to restrict the search for the mutated gene to upstream/downstream proteins.

When the mutated gene has been found (by complementation group studies) or suspected (by WB), the determination of the specific DNA mutation can be performed in a laboratory with specific instrumentation and expertise, beginning – if the gene is only suspected after WB – from the most frequent gene involved in the patient's specific population, generally the FANCA gene. If the patient comes from a population in which a founder effect has been documented, a limited number of specific mutations can be targeted. Specifically, Ashkenazi Jewish population shows FANCC and FANCD1 specific mutations (Verlander et al. 1995); non-Ashkenazi Jewish Moroccan, Afrikaner, Tunisian, Indian, Brazilian and Spanish Gypsy ethnic groups show FANCA mutations specific of each one subpopulation; Israeli Arabs display FANCA and FANCG typical mutations; sub-Saharian African Blacks own a characteristic FANCA mutation (Alter 2008).

Once a pathologic mutation has been identified in a family, a carrier testing becomes possible, searching for the mutation in the family members at risk, of both genders in most cases, or only in females at risk for FANCB.

Epidemiologic Aspects

FA affects any race and ethnic population. In the general population the prevalence is estimated at 1:100,000 live births for USA, Japan and Europe, raising significantly in some selected populations, as Ashkenazi Jews, south-African Blacks and Spanish Gypsies.

A carrier frequency of 1 in 300 is widely reported, but it should be higher (Rosenberg et al. 2011) as heterozygous subjects do not show any CI at diagnostic tests, then they may be recognized only in families where the pathologic mutations have been found in patients.

The likelihood of a carrier marrying another is calculated as about 1 in 40,000.

Although FA onset is mainly during the infancy, the disease must be suspected in adults affected by unexplained thrombocytopenia or aplastic anemia (Zatterale et al. 1995) as well as by MDS or malignancies at an unusual young age, or showing an unusual sensitivity to chemotherapy, radiotherapy or hyperbaric oxygen-therapy (Zatterale, unpublished).

FA affects equally both genders (FANCB is extremely rare). In the literature male patients are reported as characterized by an increase of CI and hematologic malignancies and a worse risk for overall survival time than females, whereas female patients display a worse hazard than males for solid tumors, due to the incidence of gynecologic cancers.

Genetic Counseling

The aim of genetic counseling is to help the patients and their families to better understand and deal with the disease causes, the therapeutic options and the psycho-social implications, in the fullest respect of their cultural and social background.

The genetic counseling issues surrounding a FA diagnosis regard inheritance and genetics, cytogenetic and molecular diagnostic tools, therapeutic options and the risk for other family members. This practice necessarily involves some different disciplines and specialists.

Regarding the reproductive counseling, a non-negligible part of the genetic counseling, prenatal testing will be offered, preferably flanking molecular to cytogenetic tests.

The parents of an affected child should be also informed that through the preimplantation genetic diagnosis technique (PGD) unaffected, HLA-compatible embryos may be obtained as future donors of HSCs to transplant in FA sibs. Ethical issues arise about such "baby design" and should be discussed with the parents of the affected child, in respect of their ethical and religious belief and by considering the social and familial context, together with technical difficulties, stressing also the emotional, physical and financial burden they will cope with.

Future

Advances in DNA sequencing techniques to detect FA mutations in a easier and more complete way are in progress.

Future hopes for FA patients relate to successful improvements in gene therapy, stem cell therapy, and stem cell gene therapy. Carcinogenesis may be seen as a major problem in FA patients, due to successful improvement in HSCT techniques and protocols and the subsequent prolonged survival of patients who after the HSCT unfortunately retain all the damages produced by the pathologic mutations they carry, except the BM failure.

More evidence is needed about the opportunity of GH therapy in short stature.

As therapy improvements are prolonging the life of FA patients to adulthood, an increasing interest and experience of adult hematologists in FA care becomes mandatory, since the transition from pediatric to adult facilities is a problem at present.

Finding natural agents or small molecules acting on the pancytopenia, as those reducing OS damage, might constitute an important therapeutic advance (Pagano et al. 2012).

The future knowledge of all the FA protein functions outside the DNA repair will lead to improved understanding of the FA pathogenesis, namely which is the primary genetic defect and which FA cellular and biological aspects are secondary phenomena, with hopeful implications in patients' expectation and quality of life.

References

Alter, B. P. (1993). Fanconi's anaemia and its variability. *Brit. J. Haemat.*, *85*, 9-14.

Alter, B. (2008). Diagnostic evaluation of FA. In: *Fanconi Anemia Guidelines for Diagnosis and Management*. Eds. Eiler, M. E., Frohnmayer, L., Owen, J. 3rd Edition. Fanconi Anemia Research Fund Inc. 33-48.

Auerbach, A. D. & Wolman, S. R. (1976). Susceptibility of Fanconi's anaemia fibroblasts to chromosome damage by carcinogens. *Nature*, *261*, 494-496.

Auerbach, A. D., Warburton, D., Bloom, A. D. & Chaganti, R. S. (1979). Preliminary communication: prenatal detection of the Fanconi anemia gene by cytogenetic methods. *Am. J. Hum. Genet.*, *31*, 77-81.

Auerbach, A. D. & Allen, R. G. (1991). Leukemia and preleukemia in Fanconi anemia patients: a review of the literature and report of the International Fanconi Anemia Registry. *Cancer Genet. Cytogenet.*, *51*, 1-12

Auerbach, A. D. (1993). Fanconi anemia diagnosis and the diepoxybutane (DEB) test. *Exp. Hematol.*, *21*, 731-733.

Barrett, A. J., Brigden, W. D., Hobbs, J. R., Hugh-Jones, K., Humble, J. G., James, D. C. O., Retsas, S., Roberts, T. R. F., Selwyn, S., Sneath, P. & Watson, J. G. (1977). Successful bone marrow transplant for Fanconi's anaemia. *Brit. Med. J.*, *1*, 420-421.

Bloom, G. E., Warner, S., Gerald, P. S. & Diamond, L. K. (1966). Chromosome abnormalities in constitutional aplastic anemia. *New Eng. J. Med.*, *274*, 8-14.

Broxmeyer, H. E., Douglas, G. W., Hangoc, G., Cooper, S., Bard, J., English, D., Arny, M., Thomas, L. & Boyse, E. A. (1989). Human umbilical cord blood as a potential source of transplantable hematopoietic stem/progenitor cells. *Proc. Nat. Acad. Sci. U S A*, *86*, 3828-3832.

Butturini, A., Gale, R. P., Verlander, P. C., Adler-Brecher, B., Gillio, A. P. & Auerbach, A. D. (1994). Hematologic abnormalities in Fanconi anemia: an International Fanconi Anemia Registry study. *Blood*, *84*, 1650-1655.

D'Andrea, A. D. (2010). Susceptibility pathways in Fanconi's anemia and breast cancer. *New Engl. J. Med.*, *362*, 1909-1919.

Davies, S. M., Radloff, G. A., DeFor, T. E., Levran, O., Batish, S. D., Hanenberg, H. & Auerbach, A. D. (2005). GST genotype may modify clinical phenotype in patients with Fanconi anemia. *Brit. J. Haematol.*, *131*, 118-122.

Demuth, I., Wlodarski, M., Tipping, A. J., Morgan, N. V., de Winter, J. P., Thiel, M., Gräsl, S., Schindler, D., D'Andrea, A. D., Altay, C., Kayserili, H., Zatterale, A., Kunze, J., Ebell, W., Mathew, C. G., Joenje, H., Sperling, K. & Digweed, M. (2000). Spectrum of mutations in the Fanconi anaemia group G gene, FANCG/XRCC9. *Eur. J. Hum. Genet.*, *8*, 861-868.

Du, W., Adam, Z., Rani, R., Zhang, X. & Pang, Q. (2008). Oxidative stress in Fanconi anemia hematopoiesis and disease progression. *Antioxid. Redox Signal.*, *10*, 1909-1921.

Dutrillaux, B., Aurias, A., Dutrillaux, A. M., Buriot, D. & Prieur, M. (1982). The cell cycle of lymphocytes in Fanconi anemia. *Hum. Genet.*, *62*, 327-332.

Estren, S. & Dameshek, W. (1947). Familial hypoplastic anemia of childhood: report of eight cases in two families with beneficial effect of splenectomy in one case. *Am. J. Dis. Child.*, *73*, 671-687.

Faivre, L., Guardiola, P., Lewis, C., Dokal, I., Ebell, W., Zatterale, A., Altay, C., Poole, J. E., Stones, D., Kwee, M. L., Van Weel-Sipman, M., Havenga, C., Morgan, N. V., de Winter, J. P., Digweed, M., Savoia, A., Pronck, J. C., De Ravel, T. J., Jansen, S., Joenie, H., Gluckman, E. & Mathew, C. G. (2000). Association of complementation group and mutation type with clinical outcome in Fanconi anemia. *Blood*, *96*, 4064-4070.

Fanconi, G. (1927). Familiäre infantile perniziosaartige Anämie (perniziöses Blutbild und Konstitution). *Jahrb. Kinderheilkl.*, *117*, 257-280.

Giri, N., Batista, D. L., Alter, B. P. & Stratakis, C. A. (2007). Endocrine abnormalities in patients with Fanconi anemia. *J. Clin. Endocrinol. Metab.*, *92*, 2624-2631.

Hussain, S., Wilson, J. B., Medhurst, A. L., Hejna, J., Witt, E., Ananth, S., Davies, A., Masson, J. Y., Moses, R., West, S. C., de Winter, J. P., Ashworth, A., Jones, N. J. & Mathew, C. G. (2004). Direct interaction of FANCD2 with BRCA2 in DNA damage response pathways. *Hum. Mol. Genet.*, *13*, 1241-1248.

Joenje, H., Arwert, F., Eriksson, A. W., de Koning, H. & Oostra, A. B. (1981). Oxygen-dependence of chromosomal aberrations in Fanconi's anaemia. *Nature*, *290*, 142-143.

Joenje, H. & Oostra, A. B. (1983). Effect of oxygen tension on chromosomal aberrations in Fanconi anaemia. *Hum. Genet.*, *65*, 99-101.

Joo, W., Xu, G., Persky, N. S., Smogorzewska, A., Rudge, D. G., Buzovetsky, O., Elledge, S. J. & Pavletich, N. P. (2011). Structure of the FANCI-FANCD2 complex: insights into the Fanconi anemia DNA repair pathway. *Science*, *333*, 312-316.

Kutler, D. I., Singh, B., Satagopan, J., Batish, S. D., Berwick, M., Giampietro, P. F., Hanenberg, H. & Auerbach, A. D. (2003). A 20-year perspective on the International Fanconi Anemia Registry (IFAR). *Blood*, *15*, 1249-56.

Levitus, M., Waisfisz, Q., Godthelp, B. C., de Vries, Y., Hussain, S., Wiegant, W. W., Elghalbzouri-Maghrani, E., Steltenpool, J., Rooimans, M. A., Pals, G., Arwert, F., Mathew, C. G., Zdzienicka, M. Z., Hiom, K., De Winter, J. P. & Joenje, H. (2005). The DNA helicase BRIP1 is defective in Fanconi anemia complementation group J. *Nat. Genet.*, *37*, 934-5.

Litman, R., Peng, M., Jin, Z., Zhang, F., Zhang, J., Powell, S., Andreassen, P. R. & Cantor, S. B. (2005). BACH1 is critical for homologous recombination and appears to be the Fanconi anemia gene product FANCJ. *Cancer Cell*, *8*, 255-65.

Liu, T., Ghosal, G., Yuan, J., Chen, J. & Huang, J. (2010). FAN1 acts with FANCI-FANCD2 to promote DNA interstrand cross-link repair. *Science*, *329*, 693-696.

Lo Ten Foe, J. R., Kwee, M. L., Rooimans, M. A., Oostra, A. B., Veerman, A. J., van Weel, M., Pauli, R. M., Shahidi, N. T., Dokal, I., Roberts, I., Altay, C., Gluckman, E., Gibson, R. A., Mathew, C. G., Arwert, F. & Joenje, H. (1997). Somatic mosaicism in Fanconi anemia: molecular basis and clinical significance. *Eur. J. Hum. Genet.*, *5*, 137-48.

Meindl, A., Hellebrand, H., Wiek, C., Erven, V., Wappenschmidt, B., Niederacher, D., Freund, M., Lichtner, P., Hartmann, L., Schaal, H., Ramser, J., Honisch, E., Kubisch, C., Wichmann, H. E., Kast, K., Deissler, H., Engel, C., Müller-Myhsok, B., Neveling, K., Kiechle, M., Mathew, C. G., Schindler, D., Schmutzler, R. K. & Hanenberg, H. (2010). Germline mutations in breast and ovarian cancer pedigrees establish RAD51C as a human cancer susceptibility gene. *Nat. Genet.*, *42*, 410-4.

Nordenson, I. (1977). Effect of superoxide dismutase and catalase on spontaneously occurring chromosome breaks in patients with Fanconi's anemia. *Hereditas*, *86*, 147-150.

Nowell, P., Bergman, G., Besa, E., Wilmoth, D. & Emanuel, B. (1984). Progressive preleukemia with a chromosomally abnormal clone in a kindred with the Estren-Dameshek variant of Fanconi's anemia. *Blood*, *64*, 1135-8.

Pagano, G., Talamanca, A. A., Castello, G., Pallardó, F. V., Zatterale, A. & Degan, P. (2012). Oxidative stress in Fanconi anaemia: from cells and molecules towards prospects in clinical management. *Biol. Chem.*, *393*, 11-21.

Pagano, G., Talamanca, A. A., Castello, G., d'Ischia, M., Pallardó, F. V., Petrović, S., Porto, B., Tiano, L. & Zatterale, A. (2013). From clinical description, to in vitro and animal studies, and backward to patients: oxidative stress and mitochondrial dysfunction in Fanconi anemia. *Free Radic. Biol. Med.*, *58*, 118-25.

Quentin, S., Cuccuini, W., Ceccaldi, R., Nibourel, O., Pondarre, C., Pagès, M. P., Vasquez, N., Dubois d'Enghien, C., Larghero, J., Peffault de Latour, R., Rocha, V., Dalle, J. H., Schneider, P., Michallet, M., Michel, G., Baruchel, A., Sigaux, F., Gluckman, E., Leblanc, T., Stoppa-Lyonnet, D., Preudhomme, C., Socié, G. & Soulier, J. (2011). Myelodysplasia and leukemia of Fanconi anemia are associated with a specific pattern of genomic abnormalities that includes cryptic RUNX1/AML1 lesions. *Blood*, *14*, 161-70.

Reid, S., Schindler, D., Hanenberg, H., Barker, K., Hanks, S., Kalb, R., Neveling, K., Kelly, P., Seal, S., Freund, M., Wurm, M., Batish, S. D., Lach, F. P., Yetgin, S., Neitzel, H., Ariffin, H., Tischkowitz, M., Mathew, C. G., Auerbach, A. D. & Rahman, N. (2007). Biallelic mutations in PALB2 cause Fanconi anemia subtype FA-N and predispose to childhood cancer. *Nat. Genet.*, *39*, 162-4.

Rosenberg, P. S., Greene, M. H. & Alter, B. P. (2003). Cancer incidence in persons with Fanconi anemia. *Blood*, *101*, 822-826.

Rosenberg, P. S., Tamary, H. & Alter, B. P. (2011). How high are carrier frequencies of rare recessive syndromes? Contemporary estimates for Fanconi Anemia in the United States and Israel. *Am. J. Med. Genet.*, *155A*, 1877-83.

Sasaki, M. S. & Tonomura, A. (1973). A high susceptibility of Fanconi's anemia to chromosome breakage by DNA cross-linking agents. *Cancer Res.*, *33*, 1829-36.

Schroeder, T. M., Anschütz, F. & Knopp, A. (1964). Spontaneous chromosome aberrations in familial panmyelopathy. *Humangenetik*, *1*, 194-6.

Sims, A. E., Spiteri, E. & Sims, R. J. (2007). FANCI is a second monoubiquitinated member of the Fanconi anemia pathway. *Nat. Struct. Mol. Biol.*, *14*, 564-567.

van der Lelij, P., Oostra, A. B., Rooimans, M. A., Joenje, H. & de Winter, J. P. (2010). Diagnostic overlap between Fanconi anemia and the Cohesinopathies: Roberts Syndrome and Warsaw Breakage Syndrome. *Anemia*, *2010*, 565268.

van Zeeburg, H. J., Snijders, P. J., Joenje, H. & Brakenhoff, R. H. (2004). Human papillomavirus DNA and p53 polymorphisms in squamous cell carcinomas from Fanconi anemia patients. *J. Natl. Cancer Inst.*, *96*, 968.

Vaz, F., Hanenberg, H., Schuster, B., Barker, K., Wiek, C., Erven, V., Neveling, K., Endt, D., Kesterton, I., Autore, F., Fraternali, F., Freund, M., Hartmann, L., Grimwade, D., Roberts, R. G., Schaal, H., Mohammed, S., Rahman, N., Schindler, D. & Mathew, C. G. (2010). Mutation of the RAD51C gene in a Fanconi anemia-like disorder. *Nat. Genet.*, *42*, 406-9.

Verlander, P. C., Kaporis, A., Liu, Q., Zhang, Q., Seligsohn, U. & Auerbach, A D. (1995). Carrier frequency of the IVS4 + 4 A-->T mutation of the Fanconi anemia gene FAC in the Ashkenazi Jewish population. *Blood*, *86*, 4034-8.

Wajnrajch, M. P., Gertner, J. M., Huma, Z., Popovic, J., Lin, K., Verlander, P. C., Batish, S. D., Giampietro, P. F., Davis, J. G., New, M. I. & Auerbach, A. D. (2001). Evaluation of growth and hormonal status in patients referred to the International Fanconi Anemia Registry. *Pediatrics*, *107*, 744-54.

Wang, X., Andreassen, P. R. & D'Andrea, A. D. (2004). Functional interaction of monoubiquitinated FANCD2 and BRCA2/FANCD1 in chromatin. *Mol. Cell. Biol.*, *24*, 5850-5862.

Zaizov, R., Matoth, Y. & Mamon, Z. (1969). Familial aplastic anaemia without congenital malformations. *Acta Paediatr. Scand.*, *58*, 151-6.

Zatterale, A., Calzone, R., Renda, S., Catalano, L., Selleri, C., Notaro, R. & Rotoli, B. (1995). Identification and treatment of late onset Fanconi Anemia. *Haematologica*, *80*, 535-538.

Zatterale, A., Kelly, F. J., Korkina, L. G., Malorni, W., Masella, R., Monti, D. & Pagano, G. (1999). Oxidative stress in cancer prone genetic diseases: a review. *Ann. Ist. Super. Sanità*, *35*, 205-209.

In: Fanconi Anemia and Oxidative Stress
Editor: Giovanni Pagano
ISBN: 978-1-63482-297-8

Chapter 2

Chromosomal Instability in Fanconi Anemia: Can It Be Mitigated by Antioxidant Small Molecules?

***Beatriz Porto*[1,*] *and Filipa Ponte*[2]**

[1]Laboratory of Cytogenetics, ICBAS- Biomedical Institute Abel Salazar, University of Porto, Portugal

[2]REQUIMTE, Laboratory of Toxicology, Faculty of Pharmacy, University of Porto, Portugal

Abstract

Fanconi Anemia (FA) is a heterogeneous genetic disorder, with 16 genes so far characterized that function in a common pathway for the maintenance of genomic stability. The clinical manifestations are also heterogeneous, although all patients share progression to bone marrow failure (BMF) and increased predisposition to cancer, particularly acute myeloid leukemia. At the cellular level chromosome instability (CI) is the hallmark of FA, and hypersensitivity of FA cells to the clastogenic effect of DNA crosslinking agents, such as diepoxybutane (DEB), provides a specific diagnostic marker.

FA cells are characterized by abnormal accumulation of reactive oxygen species (ROS) and dysfunctional response to oxidative stress (OS). They are in a permanent pro-oxidant state, demonstrated by oxidative DNA damage, increased lipid peroxidation, free iron levels, ROS overproduction, mitochondrial dysfunction, and glutathione (GSH) depletion. The hypersensitivity of FA cells to DEB is a consequence of its OS-related mechanism of cytotoxicity. DEB induces oxidative damage by forming DNA-DNA and DNA-protein crosslinks, associated with GSH depletion and activated mitochondrial apoptotic pathway.

The progressive BMF and the increased risk of malignancy associated with FA highlights the importance of understanding how the mechanisms involved in cellular

*Corresponding author: E-mail: bporto@icbas.up.pt.

defense against DNA damage fail. It was also shown that dysfunction of FA proteins increases OS damage by down-regulation of antioxidant defense genes. Thus, OS-related CI should be counteracted by other sources of antioxidants, independently of the FA pathway.

Recently, it was demonstrated that a cocktail with N-acetylcysteine, a GSH repletor and α-lipoic acid, a mitochondrial nutrient, improved genetic stability *in vitro*, decreasing CI in cultured lymphocytes from FA patients. These two active thiol antioxidants have already been used in many diseases as pro-glutathione dietary supplements. The clinical relevance of this study is suggested, considering the potential use of small molecules as a prophylactic approach to improve the defense against OS-induced cell damage and, consequently, to delay the clinical symptoms associated with this damage.

Keywords: Fanconi Anemia, chromosome instability, oxidative stress, antioxidant defense, small antioxidant molecules

Introduction

Fanconi Anemia (FA) is a highly complex and heterogeneous genetic disorder, but pivotal features are common to all patients. At molecular level, the FA pathway functions as a whole for the maintenance of genomic stability. At clinical level, the most important common manifestations that define the severity of FA disease are progressive bone marrow failure (BMF), usually arising early in life and leading to transfusion dependence, and increased predisposition to cancer. At cellular level, FA cells share increased levels of chromosome instability (CI), abnormal accumulation of reactive oxygen species (ROS) and dysfunctional response to oxidative stress (OS). In fact, OS-induced DNA damage is one of the major sources of genomic instability, which leads to increased apoptotic hematopoiesis and to BMF (Alter and Kupfer 2002).

There is no cure to date for the FA genetic defects. Therefore, BMF is currently treated with hematopoietic stem cells transplantation (HSCT), progression of pancytopenia can be delayed with androgen administration and severe neutropenia can be improved with cytokine treatments. However, there are no established preventive therapies against OS-induced damage, which could delay the progression towards BMF.

This chapter will focus on central issues relating CI with OS in FA cells and its importance for clinical management. Starting from a historic perspective to the present knowledge and toward future clinical perspectives, the following questions will be addressed: 1) CI as the hallmark of FA and DEB-induced CI as a gold standard test for the diagnosis; 2) DEB cytotoxicity and detoxification in FA cells; 3) pro-oxidant state of FA cells: a direct cause for CI increment; 4) antioxidant agents as protectors against CI in FA cells, and 5) FA treatment: current therapies and the promising role of selected antioxidant small molecules.

CHOMOSOME INSTABILITY: THE CELLULAR HALLMARK OF FANCONI ANEMIA

Diagnosis of Fanconi Anemia: From Clinical to Cytogenetic Markers

Guido Fanconi (Fanconi 1927) first described a familial form of aplastic anemia in three siblings aged 5 to 7 years with short stature, hypogonadism, skin pigmentation, hemochromatosis and pancytopenia. This syndrome was later named as Fanconi Anemia (FA) (Levy 1952). Since then new cases were documented, with the affected individuals presenting a variable range of symptoms, including progressive BMF, pigment anomalies of the skin, skeletal abnormalities, especially of the thumb and radius, kidney malformations, low birth weight, short stature, microcephaly and hypogonadism. During many years the diagnosis of FA was based only on the described clinical manifestations. However, the phenotype was so variable that an accurate evaluation was difficult and many cases remained undetected or misdiagnosed.

Since 1964 and through 1970s CI appeared as an important cellular abnormality associated with the FA phenotype. Schroeder and co-workers showed a significant increase of CI in cultured peripheral blood lymphocytes and dermal fibroblasts from FA patients comparatively to cells from normal individuals (Schroeder et al. 1964). Few years later, they suggested the use of spontaneous CI as a cellular marker for the FA diagnosis (Schroeder et al. 1976) and other reports supported this diagnostic tool (Bloom et al. 1966; Swift and Hirschhorn 1966). However, subsequent studies showed that some patients with clinical FA features had no increased spontaneous CI, and that CI frequencies appeared to be variable, ranging from none to high levels (Bushkell et al. 1976; Schroeder et al. 1976). Thus, the use of spontaneous CI as a diagnostic tool could not be established with certainty.

DEB-Induced Chromosome Instability: The Gold Standard Test for Fanconi Anemia Diagnosis

The FA diagnosis gained accuracy when it became evident that FA cells are sensitive to the clastogenic effect of some crosslink alkylating agents, such as mitomycin C (MMC) and 1,2:3,4-diepoxybutane (DEB) (Auerbach and Wolman 1976). Based on this evidence, Auerbach and co-workers (1981) reported a cytogenetic method for detection of CI in cultured lymphocytes exposed to nontoxic DEB concentrations. They showed that FA lymphocytes exhibited a 175-fold increase in CI compared to control lymphocytes. The most characteristic feature of the induced breakage was the presence of complex rearrangements involving three, four or more chromosomes, namely tri- and tetra-radial figures (Figure 1). Using this cytogenetic method, it became possible to identify previously undiagnosed pre-anemic cases, as well as patients with aplastic anemia without other clinical manifestations. Importantly, the DEB-induced CI could be considered as an specific marker of FA cells, since lymphocytes from patients with disorders displaying some of the clinical features of FA were not sensitive to DEB.

In 1989 the International Fanconi Anemia Registry (IFAR) reported the results of a cytogenetic analysis performed in 310 cases from the IFAR database, and showed that the classification of patients in two groups, DEB-sensitive (DEB+) and DEB-insensitive (DEB-), was a useful and reliable discriminator method for FA (Auerbach et al. 1989). Since then, DEB-induced CI became the most widely used test in the diagnosis of FA.

It is well established that FA is a highly heterogeneous disorder, although all patients share progression to BMF which is an indirect consequence of genomic instability. Despite the molecular and clinical variability, at the cellular level CI is the hallmark of FA. Therefore, cytogenetic analysis for evaluation of DEB-induced CI is, in fact, a unique marker, considered the gold standard test for the diagnosis of FA.

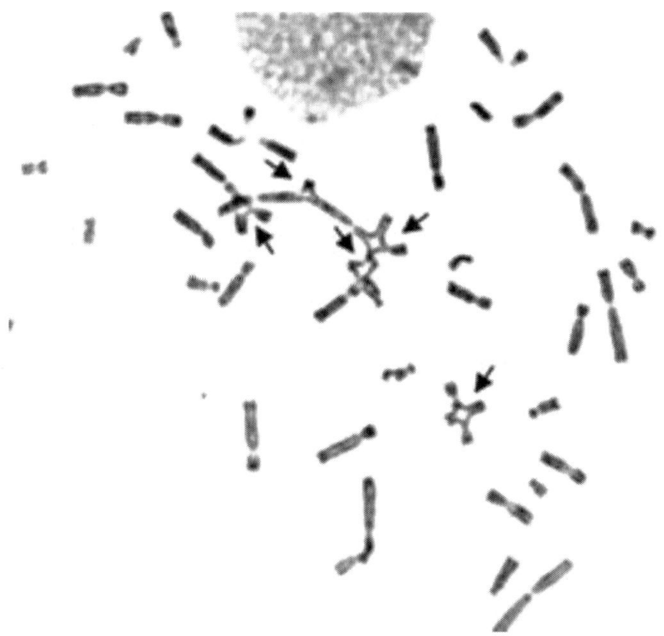

Figure 1. Diepoxybutane (DEB)-induced chromosome instability (CI) in Fanconi Anemia (FA) - in lymphocytes from a FA patient. Selected metaphase from a FA lymphocyte culture exposed to 0.1 µg/ml of DEB for 48 h. Chromosomes were stained with 4% Giemsa solution. High levels of CI can be visualized; the presence of tri- and tetra-radial figures (arrow) is particularly important for the diagnosis, since they can discriminate FA from the other CI syndromes. The image was captured with an optical microscope (500× amplification).

DEB Cytotoxicity and Detoxification in Fanconi Anemia Cells

Understanding the mechanisms involved in DEB cytotoxicity and detoxification is an important step to further understand what is failing in FA cells, which is the cause for its DEB hypersensitivity, what causes the increased levels of CI and how can be FA cells protected against CI.

DEB Cytotoxicity: Biotransformation, Crosslink Activity, Redox Impairment and Mitochondrial Dysfunction

1,2:3,4-DEB is the most toxic metabolite of 1,3-butadiene (BD). BD is not toxic by itself, but its metabolic activation forms butadiene epoxides that cause toxicity (Henderson et al. 2000). BD biotransformation produces both DEB, and two others metabolites, 3,4-epoxy-1-butene (EB) and 1,2-epoxy-3,4-butenodiol (EBD). Among these metabolites, DEB is the most toxic, 100-fold more mutagenic than EB and EBD, and also 50-fold more effective than EB in the induction of sister chromatid exchanges (SCE) in human lymphocytes (Zhang and Elfarra 2005). Therefore, DEB is considered the key metabolite responsible for BD-toxicity.

DEB toxicity is related to its epoxide structure and to its crosslinking ability; this can drastically damage DNA, by the formation of intrastrand and interstrand crosslinks (ICL), which prevent DNA unwinding, thereby blocking DNA replication, transcription and recombination (Pang and Andreassen 2009; Shen and Li 2010).

Healthy cells are capable to repair crosslinks, but in FA cells the genetic defect involving the FA pathway may account for a reduced capacity for ICL repair and, consequently, for the hypersensitivity to crosslinking agents. Nevertheless, the deficiency in DNA-DNA crosslink repair is not the only explanation for DEB cytotoxicity. It has also been related to its ability to react with protein sulfhydryl (SH) groups, producing DNA-protein crosslinks which are extremely toxic. In fact, Michaelson-Richie and co-workers (2010) showed that high concentrations of DEB can produce DNA-protein crosslinks involving one of 39 known human proteins mainly reacting with their cysteine residues. Interestingly, the cellular function of the majority of these proteins is related with transcription/translation/RNA splicing, cellular homeostasis/cell cycle and DNA replication/repair and regulation of chromatin structure.

A consistent body of evidence points to the association of DEB toxicity with redox mechanisms. First, it was described that the epoxide structure of DEB implies redox-mediated catalysis in the rearrangement of oxygen bonds (Bartók and Láng 1980). Secondly, it became widely known that GSH-based pathways are involved in the metabolism and detoxification of DEB. Korkina and co-workers (2000) showed that DEB induces high mortality and modulates catalase activity, affecting hydrogen peroxide (H_2O_2) balance in sea urchin embryos. This study supported the body of evidence associating DEB toxicity with OS. More recently, Ponte and co-workers (2011) confirmed this association in human cells. In fact, they reported that acute exposure of primary lymphocytes from normal individuals to DEB results in severe GSH depletion and loss of ATP, followed by cell death. Altogether, a number of independent studies support the relationship between hypersensitivity to DEB and OS, and emphasized that GSH-based pathways have an important role in this system.

Carcinogens, and in particular alkylating agents, have long been known to preferentially target mitochondrial DNA (mtDNA) (Wunderlich et al. 1972; Backer and Weinstein 1980; 1982; Clarke et al. 1998). The mtDNA can bind by 20- to 600-fold the amount of alkylating agents compared to genomic DNA. This difference may be, in part, due to the fact that mtDNA is not protected by histone binding, suggesting that mitochondria can be particularly sensitive to DEB-induced damage. Therefore, studies with mitochondrial protective agents may be very useful to clarify the importance of mitochondrial damage in FA phenotype, also in view of the potential interventions to prevent clinical symptoms associated with OS toxicity.

DEB Detoxification: The Protective Role of Red Blood Cells

Since DEB toxicity is redox-related and OS dependent, antioxidant systems are crucial for DEB detoxification. Among all cellular systems, red blood cells (RBCs) have shown to have a fundamental role.

RBCs are unique cells, the most abundant in the human organism, and highly specialized for their primary function, i.e., O_2 transport from lungs to tissues through hemoglobin (Hb). Consequently, they have extremely high O_2 levels which account for their high susceptibility to oxidative damage. In fact, the ferrous iron of Hb is continuously exposed to high O_2 concentrations, undergoing oxidation to methemoglobin (metHb), which is unable to bind O_2. Under OS conditions, metHb levels can reach toxic levels. In order to protect against those adverse conditions, RBCs are equipped with effective enzymatic and non-enzymatic antioxidant systems that make them mobile free radical scavengers, providing antioxidant defense both to themselves and to other tissues and organs (Siems et al. 2000; Arbos et al. 2008). Reduced GSH, ascorbic acid, α-tocopherol and other thiol groups are the major non-enzymatic antioxidants. Superoxide dismutase (SOD), glutathione peroxidase (GPx), glutathione reductase (GR) and catalase (CAT) are some of the main endogenous enzymatic defenders. All these systems are common to general aerobic cells. However, being circulating RBC some of the first cells to be affected by changes in the redox status of the body, they are widely used in first step-diagnosis of OS pathological conditions (Pandey and Rizvi 2011). Concerning FA patients, previous reports have shown RBCs impairment by deficiency in SOD activity, although other enzymes, namely GPx and CAT, may present normal activities (Joenje et al. 1979; Okahata et al. 1980; Yoshimitsu et al. 1984). Interestingly, Gille and co-workers (1987) reported that FA fibroblasts exhibited normal CuZn-SOD activities and normal levels of GSH, but Mn-SOD, CAT, and GPx activities were consistently higher. They hypothesized that the elevation of these activities might reflect a cellular pro-oxidant state in FA resulting from an increased formation of endogenous oxidizing molecular species that trigger enhanced synthesis of enzymatic antioxidant defenses (Gille et al. 1987).

The shape of the mature RBC, a biconcave disc, is also critical for its normal function. It is determined by membrane proteins, the spectrin network and the lipid bilayer. Even minimal changes in the surface may lead to morphological and functional abnormalities. Attack of ROS causes lipid peroxidation and formation of unwanted products, such as malondialdehyde (MDA) and thiobarbituric acid reactive substances (TBARS), which alter RBC shape. Straface and co-workers (2000) showed, in a morphometric study, a high rate of RBCs from FA patients significantly altered, and they hypothesized that those morphologic changes could be the result of the redox imbalance that could lead to alterations of RBC plasticity and of deformation-associated functions.

It is well established that *GSTT1* gene, expressed in RBC, is involved in determining the sensitivity of lymphocytes to DEB-induced CI (Landi, et al. 1997; 1998). However, it was also shown that DEB toxicity in primary lymphocytes can be decreased by the presence of RBCs, being this protective role independent of *GSTT1* expression and dependent on Hb content (Porto et al. 2003). Importantly, the protection was shown to be significantly increased in fetal RBC (where oxygen affinity is higher) when compared with adult RBC (Porto et al. 2006). FA is characterized by increased levels of fetal Hb; however, the clinical importance of this feature has never been directly correlated with CI. According to the described studies, it can be postulated that the increased fetal Hb concentration in FA patients may be an important additional/supportive factor involved in OS protection independently of other factors.

In order to understand if abnormalities in FA RBC could affect their antioxidant protective role, a study was performed testing the effect of normal RBC on the protection against DEB-induced CI in FA primary lymphocytes. The results clearly showed that normal RBC had a significant protective effect when compared with FA RBC (Porto et al. 2010). Although this effect was not sufficient to return DEB sensitivity to the normal range, a partial correction was evident, both *in vitro* and *in vivo*, once a significant reduction in CI was observed in primary FA lymphocytes collected after RBC transfusion (Porto et al. 2010).

Despite their high relevance, the above mentioned cellular antioxidant systems are not the exclusive participants in the defense against the pro-oxidant state of FA cells. Other important sources of antioxidant activity may play major roles as ROS scavangers, precursors of GSH synthesis and mitochondrial nutrients, as discussed further.

Pro-Oxidant State of Fanconi Anemia Cells: A Cause for Increased Chromosome Instability

The role of OS in FA cells damage was suggested first by Nordenson (1977), who showed that SOD and CAT decreased the frequency of spontaneous CI in cultured lymphocytes from FA patients. Joenje and co-workers established a direct correlation between high O_2 tension and increased damage in FA lymphocytes (Joenje et al. 1981). Two years later, Nagasawa and Little (1983) reported that SOD suppresses the cytotoxic effects of MMC in FA fibroblast cultures, and proposed that the hypersensitivity of FA cells to MMC may involve, among other factors, a deficiency in the inactivation of ROS. Importantly, Joenje and Oostra (1986) documented that lymphocytes from FA patients cultured in the presence of cyclophosphamide (CP), that is known to induce CI only after exogenous metabolic activation, exhibited excess CI even without deliberate activation of the drug, and postulated that the direct clastogenic effect of non-activated CP may be of value for the screening of possible hypersensitivity in patients considered for CP therapy. All these observations were consistent with the postulate that FA cells are in a pro-oxidant state, which can result in an increased level of spontaneous CI.

Several studies were further developed in order to better understand the pro-oxidant state of FA cells. Korkina and co-workers (1992) reported that overproduction of ROS in FA leukocytes results from activation of NADPH oxidase, decrease in the cellular antioxidant defense systems, increase in cellular free iron levels and deficiency in RBC SOD activity. Another report suggested that the high susceptibility of FA lymphoblastoid cells to oxidative DNA damage is possibly due to a decrease in CAT activity (Takeuchi and Morimoto 1993). In addition, a study by Ruppitsch and co-workers (1997) indicated that cytochrome P-450 enzymes, especially CYP1A2, play a crucial role in ROS metabolism in FA cells, that showed reduced O_2 consumption and increased activity of GPx. This is the only enzyme known to metabolize phospholipid hydroperoxides and, together with α-tocopherol, is the only known protection against ROS within the cell membrane. In fact, MDA was found to be elevated in FA cells (Ruppitsch et al. 1997). Another important marker of an *in vivo* pro-oxidant state in WBC from FA patients is the excessive accumulation of 8-hydroxy-2'-deoxyguanosine (8-OHdG), and this excess was also correlated with spontaneous CI (Degan et al. 1995; Pagano et al. 2004). Zunino and co-workers (2001) showed an increased basal level of 8-OHdG in FA lymphoblastoid lines from the complementation groups C and E, and demonstrated that H_2O_2-induced 8-OHdG formation is higher in FA than in normal cells. However, more recently Castillo and co-workers (2011) concluded that the FA pathway is redundant for the repair of this oxidative damage. Therefore, the accumulation of 8-OHdG in FA cells reflects sustained overproduction of ROS rather than defective processing of oxidized bases.

The influence of OS in the progression of cell cycle is another important feature related with FA cells. Schindler and Hoehn (1988) reported a decrease in FA fibroblast growth when cells were cultured in conditions of high O_2 tension, with accumulation of cell cycle in S and G2/M phases. However, growth of FA cells *in vitro* could be restored to near normal under hypoxic cell culture conditions. This study was later confirmed by Saito and co-workers (1993), who concluded that hypersensitivity to O_2 tension is a secondary defect in FA cells which may contribute to BMF. Poot and co-workers (1996) also tested cell cycle modulation in FA-lymphoid cell lines under various O_2 tensions, and they suggested that O_2 sensitivity might be mediated by the amount of iron in culture, which implicates a reduced capacity of FA cells to deal with ROS. Nowadays, the analysis of cell cycle progression with G2 arrest is used as an additional tool in the characterization of FA phenotype.

Link between FA Pathway and the Response to Pro-oxidant State

The involvement of the FA pathway in DNA repair/replication has already been extensively studied. But recently Du and co-workers (2012) have also shown its involvement in safeguarding the cellular antioxidant defense system. In fact, they showed that major antioxidant defense genes are down-regulated in FA patients, and this down-regulation is associated with increased DNA damage in the promoters of those genes. They proposed that one critical function of FA proteins under OS damage may be to regulate gene expression of antioxidant defense through protection of the referred promoters from oxidative DNA damage. Therefore, it is more likely that increased OS damage is due to loss of protective gene function than lack of DNA repair (Du et al. 2012). These findings may explain in part the hypersensitivity of FA cells towards O_2 and OS-related toxic agents, by means of a significant down-regulation of the genetic antioxidant control. Being so, OS may be counteracted by other enzymatic and non-enzymatic antioxidant defenses that act independently of the FA pathway genetic control. Altogether, the above mentioned studies provide both direct and indirect evidence that OS impairment involved in FA phenotype is a direct cause for the increased CI, which may be counteracted by antioxidant agents other than those dependent on the FA pathway.

Antioxidant Agents as Protectors against Chromosome Instability in Fanconi Anemia Cells

A number of *in vitro* studies have been performed testing the use of antioxidants for decreasing chromosome damage in FA cells. Dallapiccola and co-workers (1985) obtained a partial correction of CI in FA lymphocytes treated with several antioxidants, namely L-cysteine, 2-mercaptoethanol, α-mercaptopropionyl-glycine and GSH, all chemicals which are known to be donors of the SH group. Later, the same group (Porfirio et al. 1989) extended this study to desferoxamine, an iron chelator, with 50% reduction of the spontaneous chromosome breakage. Korkina and co-workers (1992) suggested that hydroxyl radical (•OH) scavengers and chelators can be used for the treatment of FA patients. Rutin, a natural bioflavonoid, was found to be an efficient inhibitor of luminol-dependent chemiluminescence (a technique that measures ROS formation) produced by FA leukocytes. After administration of rutin to three FA patients it was observed a substantial decrease in ROS production by leukocytes, a decrease in CI and improvement of hematological endpoints (Korkina et al. 1992). Pincheira and co-workers demonstrated that α-tocopherol decreases the frequency of CI and the duration of G2 in FA lymphocytes (Pincheira et al. 2001). More recently, it was reported that honey can have a cytoprotective effect against CI in FA cells, due to the antioxidant capacity of flavonoids and other polyphenols, which are common constituents of honey (Mogib El-Dahtory and Yahia 2011). The results of these studies support the assumption of O_2 sensitivity in FA cells and the putative beneficial effect of antioxidant treatments.

Some *in vivo* studies were also performed in animal models. Zhang and co-workers (2008) showed that dietary supplements with the antioxidant tempol, a nitroxide antioxidant and a SOD mimetic, can delay the age of onset of epithelial tumors in Fancd2(-/-) knock-out mice. More recently, the same group also demonstrated that the antioxidant resveratrol maintains Fancd2(-/-) KSL cells in quiescence, improves the marrow microenvironment, partially corrects the abnormal cell cycle status, and significantly improves the spleen colony-forming capacity of Fancd2(-/-) bone marrow cells (Zhang et al. 2010).

In spite of the promising results from the previous studies, no therapy using antioxidants, megavitamins, or micronutrients has been established as an effective treatment for FA patients using evidence-based criteria. Thus, further studies with new protective agents are required for the development of drugs with high effectiveness. Ideally, the protective agents/antioxidants should be already approved for human use or promptly bioavailable by dietary administration (Pagano and Korkina 2000; Pagano et al. 2012). For a rare disease like FA, insuperable obstacles to clinical chemoprevention trials may currently be faced due to the scanty numbers of patients susceptible to be recruited, compromising the best efforts in drawing adequate conclusions from small scale clinical studies.

Selected Antioxidants for Prevention against Chromosome Instability

Based on the unique hypersensitivity of FA cells to DEB, Ponte and co-workers (2012) proposed that a good choice for the use of antioxidant molecules to counteract OS damage could rely on their capacity to reduce the clastogenic effect of DEB, in order to find protective drugs that may turn out to be useful to prevent spontaneous CI. Thus, a selection among putative inhibitors of DEB bioactivation, antioxidant agents, mitochondrial protective agents, inhibitors of protective enzymes and inhibitors of protein synthesis, led to identifying the two small molecules N-acetylcysteine (NAC) and α-lipoic acid (ALA). These display protective roles against GSH depletion (Ponte et al. 2011) and, mainly, they showed an increased capacity to reduce CI in DEB-induced primary lymphocytes of both FA patients and healthy individuals (Ponte et al. 2012). Most important, for FA patients it was shown that the protective effect was significantly increased when both molecules acted in synergy, i.e., the cocktail NAC+ALA reduced significantly (between 60% and 80%) CI in spontaneous and DEB-induced FA lymphocyte cultures (Ponte et al. 2012). Assuming that the antioxidant activities leading to CI reduction *in vitro* can also occur *in vivo*, it can be hypothesized that the NAC+ALA cocktail may be a promising prophylactic treatment to decrease the adverse effects of oxidative damage in FA patients. This assumption is based on the particular properties of both antioxidants.

ALA, an endogenous and natural compound, is a mitochondrial nutrient, with a great range of action. ALA and its reduced form, dihydrolipoic acid (DHLA), form a redox couple, being DHLA the predominant form (Gorąca et al. 2011). Both are small molecules that contain two oxidized or reduced SH groups. Direct protection includes its capability to act as a potent antioxidant in redox-cycling environment, reducing vitamin C, vitamin E and coenzyme Q_{10} (Tarnopolsky 2008), as a scavenger of ROS, namely $^{\bullet}OH$, singlet oxygen ($^{1}O_2$) and hypochlorous acid (HClO) (Han et al. 1997), as a possible chelator of redox active metals (Gorąca et al. 2011) and as an essential cofactor for many mitochondrial localized enzymes (Prahalathan et al. 2006), especially mitochondrial dehydrogenases. Recently, ALA was reported to increase ALDH2 activity in OS- related diseases (Wang et al. 2011; Li et al. 2013). Indirect protection includes regulation of enzymes that act in biotransformation of xenobiotics or have an antioxidant activity, such as GST, GPx, SOD and GR (Lii et al. 2010; Odabasoglu et al. 2011). Furthermore, the ALA/DHLA couple has the ability to increase or maintain GSH levels, by acting as a transcriptional inducer of genes governing GSH synthesis and, potentially, by increasing substrate availability, mainly cysteine (Han et al. 1997; Shay et al. 2008). Consequently, ALA has the potential to maintain the antioxidant status of the cells, as demonstrated by Zembron-Lacny and co-workers (2009), protecting mtDNA from oxidation. Additionally, ALA/DHLA may reduce the proinflammatory conditions by its interaction with NF-kB, inhibiting its activity via TNF-α (Shay et al. 2008). In fact, FA cells show enhanced intracellular NFκB activation and transcriptional activity compared to normal cells (Macé et al. 2007), by impairing the immunological system. NFkB, besides being a potent indicator of stress factors such as ROS, is also a transcription factor that induces the expression of many genes involved in inflammation pathway, endothelial cell migration and apoptosis.

NAC, the acetylated variant of the amino acid L-cysteine, has been shown to increase GSH levels, the body's major antioxidant. Among the three GSH amino acids (glutamate, glycine, and cysteine), cysteine has the lowest intracellular concentration. Because *de novo* synthesis is the primary mechanism by which GSH is replenished, cysteine availability is fundamental to maintain the rate of GSH synthesis in the event of OS (Millea 2009). Moreover, NAC has the ability to reduce extracellular cystine to cysteine (Kelly 1998), due to its capacity to be a source of SH metabolites. As a source of SH groups, NAC also enhances GST activity, promotes detoxification and acts directly as a ROS scavenger. Additionally,

NAC modulates proinflammatory cytokine levels, as IL-1, TNF-α and IFN-γ (Kelly 1998), which can be useful to prevent infections in FA patients, similarly to the effects of ALA. A report by Zembron-Lacny et al. (2009) demonstrated that NAC can modulate the hematological response in healthy and active males, in particular erythropoietin (EPO), the cytokine that controls erythropoiesis. It has been shown that EPO production and erythroid differentiation are regulated by ROS, especially H_2O_2, which are involved in redox-sensitive signaling pathways through downregulation of transcription factors (Fandrey et al. 1994; Huang et al. 1996; Nagata et al. 2007). This suggests that ROS generation can suppress EPO synthesis, whereas antioxidants can stimulate its synthesis. This particularity could be very useful for the regulation of erythropoiesis in FA cells, avoiding the characteristic BMF.

Fanconi Anemia Treatment: Current and Future Therapies

As previously pointed out, the important clinical feature that defines the severity of FA disease is the progression to BMF, usually arising early in life and leading to transfusion dependence. Although there is no cure for the genetic defect, BMF can be successfully treated with HSCT, progression of pancytopenia can be delayed with androgen administration and severe neutropenia can be improved with cytokine treatments. Gene targeting is progressively becoming a realistic therapeutic alternative in clinics (Rio et al. 2014). However, it is still too early to know when this technology will be suitable for the treatment of FA patients. Therefore, gene therapy as a promising therapy will not be discussed in the present chapter.

HSCT is a curative therapy for the hematologic manifestations of FA, including progressive BMF and myeloproliferative neoplasms, although it does not cure other non-hematopoietic complications. Ideally the HSCT should be performed before multiple transfusions are needed for BMF support and definitely prior to progression from myelodysplastic syndrome to acute myeloid leukemia (AML) (MacMillan and Wagner 2010). However, issues regarding timing of transplant are complicated by the eventual difficulty in finding compatible HSC donors, the up-front risk of transplant-related mortality and the unknown long-term side effects of transplant. Since individuals with FA are extremely sensitive to the toxicity of the usual chemotherapy and radiation regimens used in preparation for HSCT, reduced doses must be used. Importantly, individuals whose hematologic manifestations have been successfully treated with HSCT appear to be at an increased risk for solid tumors, particularly head and neck squamous cell carcinomas (HNSCC). In a study conducted by Rosenberg and co-workers (2005) the risk of developed HNSCC was increased fourfold and the median age of onset was 16 years younger in transplanted *vs.* non-transplanted FA patients. In summary, HSCT is presently the only curative treatment for the hematological complications related to FA. However, late malignancies continue to be a major issue to deal in the near future, as more FA children are long time survivors of transplant.

Administration of androgens has been recommended for FA patients with BMF for whom there is still no acceptable HSCT donor. The major effects of this therapy are an increment of Hb levels and improvement of erythrocyte, platelet and neutrophil counts. Treatment is initiated when BMF is going down to clinically significant levels but before the marrow becomes completely devoid of hematopoietic stem cells, in order to permit androgens to stimulate. The mechanism(s) whereby androgens raise blood counts is currently unclear. The advantages of this treatment include the low risk of therapy-related mortality and the long history of experience with their use. The major potential side effects associated with androgen therapy include liver toxicity reflected as elevated liver enzymes, cholestasis, variable degree of virilization, peliosis hepatis and hepatic tumors (Shimamura and Alter 2010). The standard recommended androgen is oxymetholone, though danazol is now becoming a good option, once it has been proved to be well-tolerated and the side effects of virilization are less pronounced (Scheckenbach et al. 2012). About half of all treated patients will respond to androgen therapy, and a subset of those who initially respond may become refractory over time. An additional significant risk is that androgens do not prevent progression to AML that, once developed, creates a significantly higher transplant risk (Shimamura 2008). In conclusion, androgen treatment is not a definite cure for BMF but can be successfully used to delay the process towards HSCT.

A number of studies have demonstrated that granulocyte colony-stimulating factor (G-CSF) or granulocyte macrophage colony-stimulating factor (GM-CSF) can improve the neutrophil counts in FA patients (Guinan et al. 1994; Rackoff et al. 1996). Treatment with G-CSF or GM-CSF should be considered if the neutropenia is associated with recurrent or serious infections, particularly if the neutrophil counts persistently fall below $500/mm^3$ or fail to rise in response to infection. A few patients have also shown improvements in Hb or platelet counts during G-CSF or GM-CSF therapy. No comparative trials of G- *vs.* GM-CSF are available in FA patients. Hematopoietic growth factors should be administered cautiously or not at all in the setting of a clonal cytogenetic bone marrow abnormality (Alter and Kupfer 2002; Shimamura 2008), in order to prevent proliferation of the abnormal clone that may lead to cancer development.

Future Prophylactic Treatments: The Promising Role of Selected Antioxidant Small Molecules

The current therapies prevent hematopoietic failure, but no one can avoid the drastic problem of CI increment caused by OS, which is on the basis of early BMF and of the increased predisposition to either AML, before transplant, and HNSCC, after transplant. Therefore, the interest for the use of small antioxidant molecules as potential preventive drugs against OS-related cellular damage is growing.

Considering the previously described properties of both NAC and ALA, it can be hypothesized that they can be a valuable choice to prevent or delay OS-induced damage. An overall scheme for ALA and NAC action in FA cellsis described in Figure 2. These two antioxidants act at different endpoints of OS protection: ROS scavenging, GSH repletion and mitochondrial nutrition. They have already shown a synergistic protective effect *in vitro*, decreasing CI in primary FA cells, which makes them good candidates for future prophylactic interventions.

Both NAC and ALA have excellent antioxidant properties and have already been approved for human use with minimal or no adverse effects, showing an excellent safety profile. In fact, NAC and ALA have been in clinical use for more than 30 years (reviewed by Pagano et al. 2014).The use of NAC is recommended in conditions characterized by depletion of GSH or OS, as cancer, AIDS, pulmonary diseases, influenza, heart diseases, as well as in acetaminophen overdose (Kelly 1998; Atkuri et al. 2007; Millea 2009), demonstrating to be a highly effective supplement. ALA has been used in the treatment of diabetic polyneuropathies. Several evidences support a role for ALA as an OS mitigator in type 2 diabetes (Golbidi and Laher 2010), also used in the treatment of atherosclerosis, cardiovascular diseases, cataract, neurodegenerative diseases, liver diseases, AIDS and cancer (Bilska and Wlodek 2005; Gorąca et al. 2011). Besides, considering its mitochondrial activity, ALA is used for aging prevention (Arivazhagan et al. 2003).

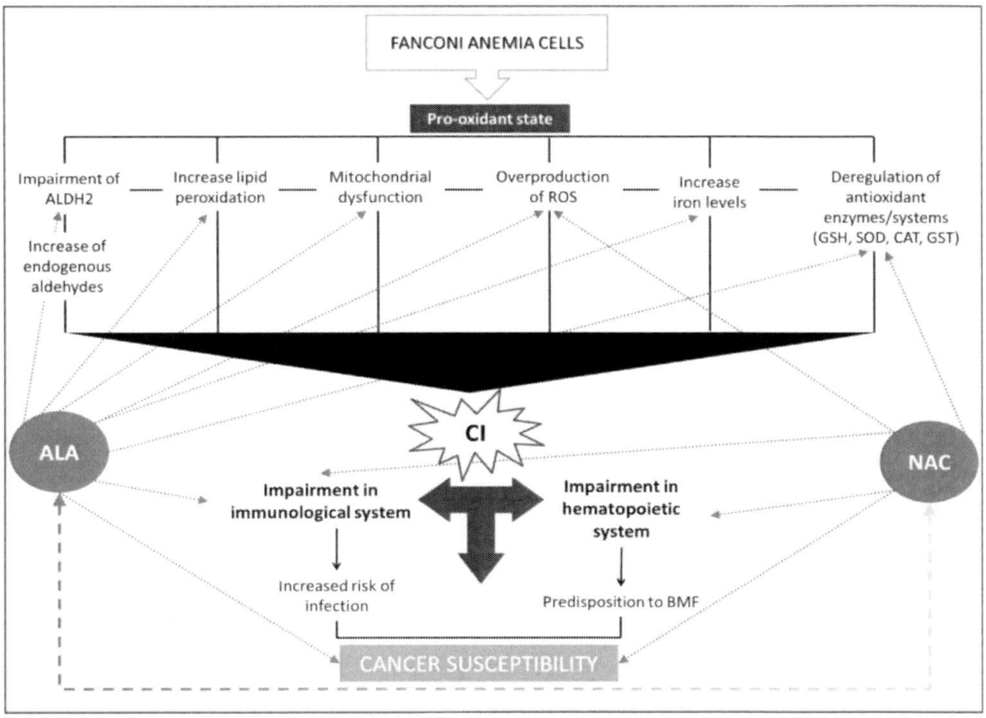

Figure 2. Proposed scheme for ALA and NAC action in FA cells. FAA cells are in permanent pro-oxidant state. These cells show increased levels of lipid peroxidation, demonstrate mitochondrial dysfunction, show increased iron levels which culminate in overproduction of ROS. The deregulation in antioxidant enzymes and systems demonstrated by FA cells contribute largely for the increment of ROS and OS. Besides, impairment of ALDH2, as a result of products of lipid peroxidation, increases the levels of endogenous aldehydes that are highly genotoxic. These characteristics result in CI, followed by impairment in immunological system and hematopoietic system, increasing the risk of infection, predisposition to BMF and cancer susceptibility. ALA and NAC can potentially prevent all these FA features. Moreover, ALA and NAC can regenerate itself together. Abbreviations: ALA, α-lipoic acid; ALDH2, aldehyde dehydrogenase 2; BMF, bone marrow failure; CAT, catalase; CI, chromosome instability; GSH, reduced glutathione; GST, glutathione S-transferase; NAC, N-acetylcysteine; ROS, reactive oxygen species; SOD, superoxide dismutase.

Both NAC and ALA show excellent pharmacokinetics. NAC is rapidly absorbed following an oral dose. Its deacetylation in the intestinal mucosa and lumen probably limits the absorption of intact molecules, however this apparently low bioavailability is misleading, since in the degradation process a variety of physiologically-beneficial, SH-containing metabolites are formed (Kelly 1998). Human organisms can synthesize ALA *de novo* from fatty acids and cysteine, but only in very small amounts. Therefore, ALA needs to be absorbed from exogenous sources, food or nutritional complements, being easily absorbed in the stomach and crossing the blood-brain barrier, due to its amphipathic properties, which allow prevention of lipid peroxidation (Bilska et al. 2005; Gorąca et al. 2011).

Additionally to CI prevention, it is also very important the fact that both NAC and ALA may modulate pro-inflammatory cytokine activities and concentrations, which opens a possibility of prevention against infections. Of note, some studies concerning the immunological response of FA patients demonstrated lymphocyte dysfunction and high levels of TNF-α (Rosselli et al. 1994). Low levels of interleukins IL-1, IL-2, and IL-6, as well as IFN-γ and granulocyte-macrophage colony stimulating factor have also been reported (Roxo et al. 2001). ALA has important cellular roles in the inhibition of the activation of NFkB and activating protein 1 (Mizuno and Packer 1994, Petersen et al. 2008), while NAC modulates cytokine levels, such as IL-1, TNF-α and IFN-γ (Kelly 1998). NAC also modulates EPO, a cytokine responsible for hematopoiesis response, as mentioned above. These pharmacological effects may further contribute to the therapeutic usefulness of NAC plus ALA cocktail by avoiding the inflammatory process, mostly frequent in FA children patients (Myers et al. 2011; Liu et al. 2012), and increasing hematopoietic response.

Conclusion

Nowadays one can hardly avoid the conclusion that FA cells have an insufficient protection against OS-related damaging effects. In view of the current lack of preventive treatments to delay clinical symptoms associated with OS in FA patients, of the fact that oral administration of selected antioxidants is well established in the treatment of various OS-related diseases with high favorable risk/benefit, and of the need to improve patients' quality of life, it can be anticipated that prophylactic interventions with antioxidant small molecules acting at different endpoints of OS protection can be of great clinical value for FA patients.

References

Alter, B. P. & Kupfer, G. (2002-updated 2013). Fanconi anemia. In: Pagon, R.A., Adam, M.P., Bird, T.D., Dolan, C.R., Fong, C.T., Stephens, K.Eds. *GeneReviews* [Internet]. University of Washington, Seattle, WA. Available at http://www.ncbi.nlm.nih.gov/books/NBK140.

Arbos, K. A., Claro, L. M., Borges, L., Santos, C. A. M. & Weffor-Santos, A. M. (2008). Human erythrocytes as a system for evaluating the antioxidant capacity of vegetable extracts. *Nutr. Res.*, *28*, 457-463.

Arivazhagan, P., Panneerselvam, S. R. & Panneerselvam, C. (2003). Effect of dl-alpha-lipoic acid on the status of lipid peroxidation and lipids in aged rats. *J. Gerontol.*, *58A*, 788-791.

Atkuri, K. R., Mantovani, J. J., Herzenberg, L. A. & Herzenberg, L. A. (2007). N-Acetylcysteine - a safe antidote for cysteine/glutathione deficiency. *Curr. Opin. Pharmacol.*, *7*, 355-359.

Auerbach, A. D. & Wolman, S. R. (1976). Susceptibility of Fanconi's Anaemia fibroblasts to chromosome damage by carcinogens. *Nature*, *261*, 494-496.

Auerbach, A. D., Adler, B. & Chaganti, R. S. (1981). Prenatal and postnatal diagnosis and carrier detection of Fanconi Anemia by a cytogenetic method. *Pediatrics*, *67*, 128-135.

Auerbach, A. D., Rogatko, A. & Schroeder-Kurth, T. M. (1989). International Fanconi Anemia Registry: Relation of clinical symptoms to diepoxybutane sensitivity. *Blood*, *73*, 391-396.

Backer, J. & Weinstein, I. (1980). Mitochondrial DNA is a major cellular target for a dihydrodiol-epoxide derivative of benzo[a]pyrene. *Science*, *209*, 297-299.

Backer, J. M. & Weinstein, I. B. (1982). Interaction of benzo(a)pyrene and its dihydrodiol-epoxide derivative with nuclear and mitochondrial DNA in C3h10t½ cell cultures. *Cancer Res.*,*42*, 2764-2769.

Bartók, M. & Láng, K. L. (1980). Oxiranes. In *The Chemistry of Functional Groups* (Vol. Supplement E, Part 2. The chemistry of ethers, crown ethers, hydroxyl groups and theirs sulfur analogues). Patai, S., Chichester, U. K. Eds. Jonh Wiley & Sons, 609-673.

Bilska, A., & Wlodek, L. (2005). Lipoic Acid - the Drug of the Future? *Pharmacol. Rep.*, *57*, 570-577.

Bloom, A. D., Neriishi, S., Kamada, N., Iseki, T. & Keehn, R. J. (1966). Cytogenetic investigation of survivors of the atomic bombings of Hiroshima and Nagasaki. *Lancet*, *2(7465)*, 672-674.

Bushkell, L. L., Kersey, J. H. & Cervenka, J. (1976). Chromosomal breaks in T and B lymphocytes in Fanconi's Anemia. *Clin. Genet.*, *9*, 583-587.

Castillo, P., Bogliolo, M. & Surralles, J. (2011). Coordinated action of the Fanconi Anemia and Ataxia Telangiectasia pathways in response to oxidative damage. *DNA Repair*, *10*, 518-525.

Clarke, A. A., Marsh, J. C., Gordon-Smith, E. C. & Rutherford, T. R. (1998). Molecular genetics and Fanconia Anaemia: New insights into old problems. *Brit. J. Haematol.*, *103*, 287-296.

Dallapiccola, B., Porfirio, B., Mokini, V., Alimena, G., Isacchi, G. & Gandini, E. (1985). Effect of Oxidants and Antioxidants on Chromosomal Breakage in Fanconi Anemia Lymphocytes. *Hum. Genet.*, *69*, 62-65.

Degan, P., Bonassi, S., De Caterina, M., Korkina, L. G., Pinto, L., Scopacasa, F., Zatterale, A., Calzone, R. & Pagano, G. (1995). *In vivo* accumulation of 8-hydroxy-2'-deoxyguanosine in DNA correlates with release of reactive oxygen species in Fanconi's Anaemia families. *Carcinogenesis*, *16*, 735-742.

Du, W., Rani, R., Sipple, J., Schick, J., Myers, K. C., Mehta, P., Andreassen, P. R., Davies, S. M. & Pang, Q. (2012). The FA pathway counteracts oxidative stress through selective protection of antioxidant defense gene promoters. *Blood*, *119*, 4142-4151.

Fanconi, G. (1927). Familiäre infantile perniziosartige Anämie (perniziöses Blutbild und Konstitution). *Jahrbuch Kinderheild*, *117*, 257-280.

Fandrey, J., Frede, S. & Jelkmann, W. (1994). Role of hydrogen peroxide in hypoxia-induced erythropoietin production. *Biochem. J.*, *303*, 507-510.

Gille, J. J., Wortelboer, H. M. & Joenje, H. (1987). Antioxidant status of Fanconi Anemia fibroblasts. *Hum. Genet.*, *77*, 28-31.

Golbidi, S. & Laher, I. (2010). Antioxidant therapy in human endocrine disorders. *Med. Sci. Monit.*, *16*, RA9-24.

Gorąca, A., Huk-Kolega, H., Piechota, A., Kleniewska, P., Ciejka, E. & Skibska, B. (2011). Lipoic acid - Biological activity and therapeutic potential. *Pharmacol. Rep.*, *63*, 849-858.

Guinan, E. C., Lopez, K. D., Huhn, R. D., Felser, J. M. & Nathan, D. G. (1994). Evaluation of granulocyte-macrophage colony-stimulating factor for treatment of pancytopenia in children with Fanconi Anemia. *J. Pediatr.*, *124*, 144-150.

Han, D., Handelman, G., Marcocci, L., Sen, C. K., Roy, S., Kobuchi, H., Tritschler, H. J., Flohé, L. & Packer, L. (1997). Lipoic acid increases de novo synthesis of cellular glutathione by improving cystine utilization. *BioFactors*, *6*, 321-338.

Henderson R. F., Barr, E. B., Belinsky, S. A., Benson, J. M., Hahn, F. F. & Ménache, M. G. (2000). 1,3-Butadiene: cancer, mutations, and adducts. Part I: Carcinogenicity of 1,2,3,4-diepoxybutane. *Res. Rep. Health Eff. Inst.*, *92*, 11-43, discussion 45-48.

Huang, L. E., Arany, Z., Livingston, D. M. & Bunn, H. F. (1996). Activation of hypoxia-inducible transcription factor depends primarily upon redox-sensitive stabilization of its α subunit. *J. Biol. Chem.*, *271*, 32253-32259.

Joenje, H., Frants, R. R., Arwert, F., de Bruin, G. J., Kostense, P. J., van de Kamp, J. J., de Koning, J. & Eriksson, A. W. (1979). Erythrocyte superoxide dismutase deficiency in Fanconi's Anaemia established by two independent methods of assay. *Scand. J. Clin. Lab. Invest.*, *39*, 759-764.

Joenje, H., Arwert, F., Eriksson, A. W., de Koning, H. & Oostra, A. B. (1981). Oxygen-dependence of chromosomal aberrations in Fanconi's Anaemia. *Nature*, *290*, 142-143.

Joenje, H. & Oostra, A. B. (1986). Clastogenicity of cyclophosphamide in Fanconi's Anemia lymphocytes without exogenous metabolic activation. *Cancer Genet. Cytogenet.*, *22*, 339-345.

Kelly, G. S. (1998). Clinical applications of N-acetylcysteine. *Altern. Med. Rev.*, *3*, 114-127.

Korkina, L. G., Samochatova, E. V., Maschan, A. A., Suslova, T. B., Cheremisina, Z. P. & Afanas'ev, I. B. (1992). Release of active oxygen radicals by leukocytes of Fanconi Anemia patients. *J. Leukoc. Biol.*, *52*, 357-362.

Korkina, L. G., Deeva, I. B., De Biase, A., Iaccarino, M., Oral, R., Warnau, M. & Pagano, G. (2000). Redox-dependent toxicity of diepoxybutane and mitomycin C in sea urchin embryogenesis. *Carcinogenesis*, *21*, 213-220.

Landi, S., Ponzanelli, I., Cipollini, G., Frenzilli, G., Luccini, A., Milillo, C. P., Sbrana, I. & Barale, R. (1997). Modulating factors of individual sensitivity to diepoxybutane: chromosome aberrations induced in vitro in human lymphocytes. *Mutagenesis*, *12*, 17-22.

Landi S., Norppa H., Frenzilli G., Cipollini G., Ponzanelli I., Baral & R, Hirvonen A. (1998). Individual sensitivity to cytogenetic effects of 1,2:3,4-diepoxybutane in cultured human lymphocytes: Influence of glutathione S-transferase M1, P1 and T1 genotypes. *Pharmacogenetics*, *8*, 461-471.

Levy, W. (1952). Aplastic anemia in siblings with multiple congenital anomalies (the Fanconi type). *J. Pediatr.*, *40*, 24-41.

Li, R. J., Ji, W. Q., Pang, J. J., Wang, J. L., Chen, Y. G. & Zhang, Y. (2013). Alpha-lipoic acid ameliorates oxidative stress by increasing aldehyde dehydrogenase-2 activity in patients with acute coronary syndrome. *Tohoku J. Exp. Med.*, *229*, 45-51.

Lii, C. K., Liu, K. L., Cheng, Y. P., Lin, A. H., Chen, H. W. & Tsai, C. W. (2010). Sulforaphane and α-lipoic acid upregulate the expression of the iiclass of glutathione S-Transferase through C-Jun and Nrf2 activation, *J. Nutr.*, *140*, 885-892.

Liu, Y., Ballman, K., Li, D., Khan, S., Derr-Yellin, E., Shou, W. & Haneline, L. S. (2012). Impaired function of Fanconi Anemia type C-deficient macrophages. *J. Leukoc. Biol.*, *91*, 333-340.

Macé, G., Briot, D., Guervilly, J. H. & Rosselli, F. (2007). L'anémie de Fanconi: Aspects cellulaires et moléculaires. *Pathol. Biol.*, *55*, 19-28.

MacMillan, M. L. & Wagner, J. E. (2010). Haematopoeitic cell transplantation for Fanconi Anaemia – When and how? *Brit. J. Haematol.*, *149*, 14-21.

Michaelson-Richie, E. D., Loeber, R. L., Codreanu, S. G., Ming, X., Liebler, D. C., Campbell, C. & Tretyakova, N. Y. (2010). DNA-Protein cross-linking by 1,2,3,4-diepoxybutane. *J. Proteome Res.*, *9*, 4356-4367.

Millea, P. J. (2009). N-Acetylcysteine: Multiple clinical applications. *Am. Fam. Physician*, *80*, 265-269.

Mizuno, M. & Packer, L. (1994). Effects of alpha-lipoic acid and dihydrolipoic acid on expression of proto-oncogene C-Fos. *Biochem. Biophys. Res. Commun.*, 200, 1136-1142.

Mogib El-Dahtory, F. A., & Yahia, S. (2011). Cytoprotective effect of honey against chromosomal breakage in Fanconi Anemia patients in vitro. *Indian J. Hum. Genet.*, *17*, 77-81.

Myers, K. C., Bleesing, J. J., Davies, S. M., Zhang, X., Martin, L. J., Mueller, R., Harris, R. E., Filipovich, A. H., Kovacic, M. B., Wells, S. I. & Mehta, P. A. (2011). Impaired immune function in children with Fanconi Anaemia. *Br. J. Haematol.*, *154*, 234-240.

Nagasawa, H. & Little, J. B. (1983). Suppression of cytotoxic effect of mitomycin-C by superoxide dismutase in Fanconi's Anemia and Dyskeratosis Congenita fibroblasts. *Carcinogenesis*, *4*, 795-799.

Nagata, M., Arimitsu, N., Ito, T. & Sekimizu, K. (2007). Antioxidant N-acetyl-L-cysteine inhibits erythropoietin-induced differentiation of erythroid progenitors derived from mouse fetal liver. *Cell Biol. Int.*, *31*, 252-256.

Nordenson, I. (1977). Effect of superoxide dismutase and catalase on spontaneously occurring chromosome breaks in patients with Fanconi's Anemia. *Hereditas*, *86*, 147-150.

Odabasoglu, F., Halici, Z., Aygun, H., Halici, M., Atalay, F., Cakir, A., Cadirci, E., Bayir, Y. & Suleyman, H. (2011). α-Lipoic acid has anti-inflammatory and anti-oxidative properties: An experimental study in rats with carrageenan-induced acute and cotton pellet-induced chronic inflammations. *Brit. J. Nutr.*, *105*, 31-43.

Okahata, S., Kobayashi, Y. & Usui, T. (1980). Erythrocyte superoxide dismutase activity in Fanconi's Anaemia. *Clin. Sci. (Lond)*, *58*, 173-175.

Pagano, G. & Korkina, L. G. (2000). Prospects for nutritional interventions in the clinical management of Fanconi Anemia, *Cancer Causes Control*, *11*, 881-889.

Pagano, G., Degan, P., d'Ischia, M., Kelly F. J., Pallardó, F. V., Zatterale, A., Anak, S. S., Akisik, E. E., Beneduce, G., Calzone, R., De Nicola, E., Dunster, C., Lloret, A., Manini, P., Nobili, B., Saviano, A., Vuttariello, E. & Warnau, M. (2004). Gender- and age-related distinctions for the in vivo prooxidant state in Fanconi Anaemia patients, *Carcinogenesis*, *25*, 1899-1909.

Pagano, G., Talamanca, A. A., Castello, G., Pallardó,F. V., Zatterale, A. & Degan, P. (2012). Oxidative stress in Fanconi Anaemia: From cells and molecules towards prospects in clinical management. *Biol. Chem.*, *393*, 11-21.

Pagano, G., Talamanca, A. A., Castello, G., Cordero, M. D., d'Ischia, M., Gadaleta, M. N., Pallardó, F. V., Petrović, S., Tiano, L. & Zatterale, A. (2014). Current experience in testing mitochondrial nutrients in disorders featuring oxidative stress and mitochondrial dysfunction: rational design of chemoprevention trials. *Int. J. Mol. Sci.*, *15*, 20169-20208.

Pandey, K. B. & Rizvi, S. I. (2011). Biomarkers of oxidative stress in red blood cells. *Biomed. Pap. Med. Fac. Univ. Palacky Olomouc Czech Repub.*, *155*, 131-136.

Pang, Q. & Andreassen, P. R. (2009). Fanconi Anemia proteins and endogenous stresses. *Mutat. Res.*, *668*, 42-53.

Petersen Shay, K., Moreau, R. F., Smith, E. J. & Hagen, T. M. (2008). Is alpha-lipoic acid a scavenger of reactive oxygen species in vivo? Evidence for its initiation of stress signaling pathways that promote endogenous antioxidant capacity. *IUBMB Life*, *60*, 362-367.

Pincheira, J., Bravo, M., Santos, M. J., de la Torre, C. & Lópe-Sáez, J. F. (2001). Fanconi anemia lymphocytes: effect of dl-α-tocopherol (Vitamin E) on chromatid breaks and on G2 repair efficiency. *Mutat. Res.*, *461*, 265-271.

Ponte, F., Carvalho, F. & Porto, B. (2011). Protective effect of acetyl-l-carnitine and alpha-lipoic acid against the acute toxicity of diepoxybutane to human lymphocytes. *Toxicology*, *289*, 52-58.

Ponte, F., Sousa, R., Fernandes, A. P., Gonçalves, C., Barbot, J., Carvalho, F. & Porto, B. (2012). Improvement of genetic stability in lymphocytes from Fanconi Anemia patients through the combined effect of alpha-lipoic acid and N-acetylcysteine. *Orphanet J. Rare Dis.*, *7*, 28.

Poot, M., Gross, O., Epe, B., Pflaum, M. & Hoehn, H. (1996). Cell cycle defect in connection with oxygen and iron sensitivity in Fanconi Anemia lymphoblastoid cells. *Exp. Cell Res.*, *222*, 262-268.

Porfirio, B., Ambroso, G., Giannella, G., Isacchi, G. & Dallapiccola, B. (1989). Partial correction of chromosome instability in Fanconi Anemia by desferrioxamine. *Hum. Genet.*, *83*, 49-51.

Porto, B., Chiecchio, L., Gaspar, J., Faber, A., Pinho, L., Rueff, J. & Malheiro, I. (2003). Role of haemoglobin in the protection of cultured lymphocytes against diepoxybutane (DEB), assessed by in vitro induced chromosome breakage. *Mutat. Res.*, *536*, 61-67.

Porto, B., Oliveira, R. J., Sousa, C., Gaspar, J., Rueff, J., Carvalho, F. & Malheiro, I. (2006). The role of foetal red blood cells in protecting cultured lymphocytes against diepoxybutane-induced chromosome breaks. *Mutat. Res.*, *603*, 41-47.

Porto, B., Sousa, R., Malheiro, I., Gaspar, J., Rueff, J., Gonçalves, C. & Barbot, J. (2010). Normal red blood cells partially decrease diepoxybutane-induced chromosome breakage in cultured lymphocytes from Fanconi Anaemia patients. *Cell Prolif.*, *43*, 573-578.

Prahalathan, C., Selvakumar, E., Varalakshmi, P., Kumarasamy, P. & Saravanan, R. (2006). Salubrious effects of lipoic acid against adriamycin-induced clastogenesis and apoptosis in Wistar rat bone marrow cells. *Toxicology*, *222*, 225-232.

Rackoff, W. R., Orazi, A., Robinson, C. A., Cooper, R. J., Alter, B. P., Freedman, M. H., Harris, R. E. & Williams, D. A. (1996). Prolonged administration of granulocyte colony-stimulating factor (Filgrastim) to patients with Fanconi Anemia: A pilot study. *Blood, 88*, 1588-1593.

Rio, P., Baños, R., Lombardo, A., Quintana-Bustamante, O., Alvarez, L., Garate, Z., Genovese, P., Almarza, E., Valeri, A., Díez, B., Navarro, S., Torres, Y., Trujillo, J. P., Murillas, R., Segovia, J. C., Samper, E., Surralles, J., Gregory, P. D., Holmes, M. C., Naldini, L. & Bueren, J. A. (2014). Targeted gene therapy and cell reprogramming in Fanconi Anemia. *EMBO Mol. Med.*, *6*, 835-848.

Rosenberg, P. S., Socié, G., Alter, B. P. & Gluckman, E. (2005). Risk of head and neck squamous cell cancer and death in patients with Fanconi Anemia who did and did not receive transplants. *Blood*, *105*, 67-73.

Rosselli, F., Sanceau, J., Gluckman, E., Wietzerbin, J. & Moustacchi, E. (1994). Abnormal lymphokine production: A novel feature of the genetic disease Fanconi Anemia. II. In Vitro and in vivo spontaneous overproduction of Tumor Necrosis Factor alpha. *Blood*, *83*, 1216-1225.

Roxo, P., Jr. Arruda, L. K., Nagao, A. T., Carneiro-Sampaio, M. M. & Ferriani, V. P. (2001). Allergic and immunologic parameters in patients with Fanconi's Anemia. *Int. Arch. Allergy Immunol.*, *125*, 349-355.

Ruppitsch, W., Meisslitzer, C., Weirich-Schwaiger, H., Klocker, H., Scheidereit, C., Schweiger, M., & Hirsc-Kauffmann, M. (1997). The role of oxygen metabolism for the pathological phenotype of Fanconi Anemia. *Hum. Genet.*, *99*, 710-719.

Saito, H., Hammond, A. T. & Moses, R. E. (1993). Hypersensitivity to oxygen is a uniform and secondary defect in fanconi anemia cells. *Mutat. Res.*, *294*, 255-262.

Scheckenbach, K., Morgan, M., Filger-Brillinger, J., Sandmann, M., Strimling, B., Scheurlen, W., Schindler, D., Göbel, U. & Hanenberg, H. (2012). Treatment of the bone marrow failure in Fanconi Anemia patients with danazol. *Blood Cells Mol. Dis.*, *48*, 128-131.

Schindler, D. & Hoehn, H. (1988). Fanconi Anemia mutation causes cellular susceptibility to ambient oxygen. *Am. J. Hum. Genet.*, *43*, 429-435.

Schroeder, T., Anschultz, F. & Knoff, A. (1964). Spontane Chromosomenaberrationen bei familiärer Panmyelopathie. *Humangenetik*, *1*, 194-196.

Schroeder, T., Tilgen, D., Krüger, J. & Vogel, F. (1976). Formal genetics of Fanconi's Anemia. *Hum. Genet.*, *32*, 257-288.

Shay, K., Moreau, R. F., Smith, E. J. & Hagen, T. M. (2008). Is α-lipoic acid a scavenger of reactive oxygen species in vivo? Evidence for its initiation of stress signaling pathways that promote endogenous antioxidant capacity. *IUBMB Life*, *60*, 362-367.

Shen, X. & Li, L. (2010). Mutagenic repair of DNA interstrand crosslinks. *Environ. Mol. Mutagen.*, *51*, 493-499.

Shimamura, A. (2008). Treatment of Hematologic Abnormalities in Fanconi Anemia. In: *Fanconi Anemia Guidelines for the Diagnosis and Management*. Eds. Eiler, M., Frohnmayer, D., Frohnmayer, L., Larsen, K., Owen, J. 3rd Edition. Fanconi Anemia Research Fund Inc. 49-75.

Shimamura, A. & Alter, B. P. (2010). Pathophysiology and management of inherited bone marrow failure syndromes, *Blood Rev.*, *24*, 101-122.

Siems, W. G., Sommerburg, O. & Grune, T. (2000). Erythrocyte free radical and energy metabolism. *Clin. Nephrol.*, *53*, S9-17.

Straface, E., Masella, R., Del Principe, D., Franceschi, C., Korkina, L. G., Zatterale, A., Pagano, G. & Malorni, W. (2000). Spectrin changes occur in erythrocytes from patients with Fanconi's Anemia and their parents. *Biochem. Biophys. Res. Commun.*, *273*, 899-901.

Swift, M. R. & Hirschhorn, K. (1966). Fanconi's Anemia inherited susceptibility to chromosome breakage in various tissues. *Ann. Int. Med.*, *65*, 496-503.

Takeuchi, T. & Morimoto, K. (1993). Increased formation of 8-hydroxydeoxy-guanosine, an oxidative DNA damage, in lymphoblasts from Fanconi's Anemia patients due to possible catalase deficiency. *Carcinogenesis*, *14*, 1115-1120.

Tarnopolsky, M. A. (2008). The mitochondrial cocktail: Rationale for combined nutraceutical therapy in mitochondrial cytopathies. *Adv. Drug Del. Rev.*, *60*, 1561-1567.

Wang, J., Wang, H., Hao, P., Xue, L., Wei, S., Zhang, Y. & Chen, Y. (2011). Inhibition of aldehyde dehydrogenase 2 by oxidative stress is associated with cardiac dysfunction in diabetic rats. *Mol. Med.*, *17*, 172-179.

Wunderlich, V., Tetzlaff, I. & Graffi, A. (1972). Studies on nitrosodimethylamine: Preferential methylation of mitochondrial DNA in rats and hamsters, *Chem. Biol. Interact.*, *4*, 81-89.

Yoshimitsu, K., Kobayashi, Y. & Usui, T. (1984). Decreased superoxide dismutase activity of erythrocytes and leukocytes in Fanconi's Anemia. *Acta Haematol.*, *72*, 208-210.

Zembron-Lacny, A., Slowinska-Lisowska, M., Szygula, Z., Witkowski, K. & Szyszka, K. (2009). The comparison of antioxidant and hematological properties of N-acetylcysteine and alpha-lipoic acid in physically active males. *Physiol. Res.*, *58*, 855-861.

Zhang, Q. S., Eaton, L., Snyder, E. R., Houghtaling, S., Mitchell, J. B., Finegold, M., Van Waes, C. & Grompe, M. (2008). Tempol protects against oxidative damage and delays epithelial tumor onset in Fanconi Anemia mice. *Cancer Res.*, *68*, 1601-1608.

Zhang, Q. S., Marquez-Loza, L., Eaton, L., Duncan, A. W., Goldman, D. C., Anur, P., Watanabe-Smith, K., Rathbun, R. K., Fleming, W. H., Bagby, G. C. & Grompe, M. (2010). Fancd2-/- mice have hematopoietic defects that can be partially corrected by resveratrol. *Blood*, *116*, 5140-5148.

Zhang, X. Y. & Elfarra, A. A. (2005). Reaction of 1,2,3,4-diepoxybutane with 2'-deoxyguanosine: Initial products and their stabilities and decomposition patterns under physiological conditions. *Chem. Res. Toxicol.*, *18*, 1316-1323.

Zunino, A., Degan, P., Vigo, T. & Abbondandolo, A. (2001). Hydrogen peroxide: Effects on DNA, chromosomes, cell cycle and apoptosis induction in Fanconi's Anemia cell lines. *Mutagenesis*, *16*, 283-288.

In: Fanconi Anemia and Oxidative Stress
Editor: Giovanni Pagano
ISBN: 978-1-63482-297-8

Chapter 3

Biochemical Grounds for "Crosslinker Sensitivity": What Have We Learned from Pharmacology?

Giovanni Pagano[*,1], *Marco d'Ischia*[2]
and Federico V. Pallardó[3]
[1]Italian National Cancer Institute, CROM, Mercogliano, Italy;
[2]Department of Chemical Sciences,
University of Naples Federico II, Naples, Naples, Italy;
[3]University of Valencia-INCLIVA, CIBERER (Centro de Investigación Biomédica en Red de Enfermedades Raras), Valencia, Spain

Abstract

The literature pointing to mitomycin C bioactivation, and to the toxicity mechanisms of diepoxybutane and a group of nitrogen mustards causing DNA crosslinks in Fanconi Anemia (FA) cells is reviewed. A critical analysis of this literature prompts revisiting the FA phenotype and crosslinker sensitivity in terms of an oxidative stress (OS) background, redox-related anomalies of FA (FANC) proteins, and mitochondrial dysfunction. This re-appraisal of FA basic defect might lead to innovative approaches both in elucidating FA phenotype and in prospect developments of patients' clinical management.

Keywords: mitomycin C, diepoxybutane, bioactivation, redox mechanisms

Introduction

Excess sensitivity to "crosslinkers", i.e., agents causing DNA interstrand crosslinks (ICL), both results in an increase of spontaneous chromosomal instability and in loss of cell

[*]Corresponding author: E-mail: gbpagano@tin.it.

viability. Fanconi Anemia cells are recognized to display excess sensitivity to crosslinkers. Thus, the shared opinion associates FA's crosslinker sensitivity to deficiency in DNA repair pathways; this opinion dates back to early reports since 1970s (Sasaki and Tonomura 1973; Fujiwara and Tatzumi 1975; Latt et al. 1975; Auerbach and Wolman 1976; Raj and Heddle 1980; Ishida R, Buchwald 1982), in analogy with other genetic diseases categorized for DNA repair defects (Fujiwara 1982). DNA repair defect is currently viewed as the shared axiom in FA definition (Kim and D'Andrea 2012; Kupfer 2013).

This chapter is to review the toxicity mechanisms of FA-related crosslinkers in view of reconsidering the main defect(s) in FA phenotype as due to abnormal response to crosslinker toxicity, as excess bioactivation pathways and/or as defective detoxification of these xenobiotics, as discussed below.

Crosslinker-Associated Toxicity Mechanisms

The early studies in 1960s of new drugs as candidate antineoplastic drugs prompted investigations on their toxicity mechanisms, including biotransformation to directly acting metabolites. This was the case for various xenobiotics resulting in DNA alkylation and, in particular, for some agents causing ICL, termed "crosslinkers". These included some anthracyclines, as mitomycin C (MMC) and some nitrogen mustards, as melphalan (MEL) and cyclophosphamide (CP). A different focus originally arose for another major xenobiotic, 1,2:3,4-diepoxybutane (DEB), a directly acting mutagenic metabolite of butadiene, utilized in the production of synthetic rubber (Norppa and Sorsa 1993).

After early reports on spontaneous chromosomal instability in lymphocytes from FA patients (Schroeder-Kurth et al. 1964), subsequent studies found that crosslinkers dramatically enhanced chromosomal instability in FA cells (Auerbach et al. 1979). This excess sensitivity to ICL-inducing clastogens, as MMC and DEB, has remained since then the key procedure in FA diagnosis.

Mitomycin C

Iyer and Szybalski (1963; 1964) first demonstrated that MMC bioactivation to its directly acting derivatives requires the addition of cell extracts. As shown in Table 1, a series of reports found that MMC bioactivation is associated with redox mechanisms involving NADPH P-450 reductase producing reactive oxygen species (ROS) with formation of semiquinone radicals (Handa and Sato 1975) (Figure 1). Other affected endpoints in MMC bioactivation included lipid peroxidation that was inhibited by antioxidant activities (superoxide dismutase, SOD and catalase, CAT), by glutathione (GSH), and by ascorbic acid (Kennedy et al. 1980; Rockwell et al. 1982; Trush et al. 1982; Gutteridge et al. 1984). MMC toxicity was found to strictly depend on hypoxia, in analogy with other crosslinkers as, e.g., MEL (Bachur et al. 1979; Teicher et al. 1981; Strese et al. 2013). As shown in Figure 1, in normoxic conditions the reduced activated form of MMC is re-oxidized and thus decreases its alkylating power.

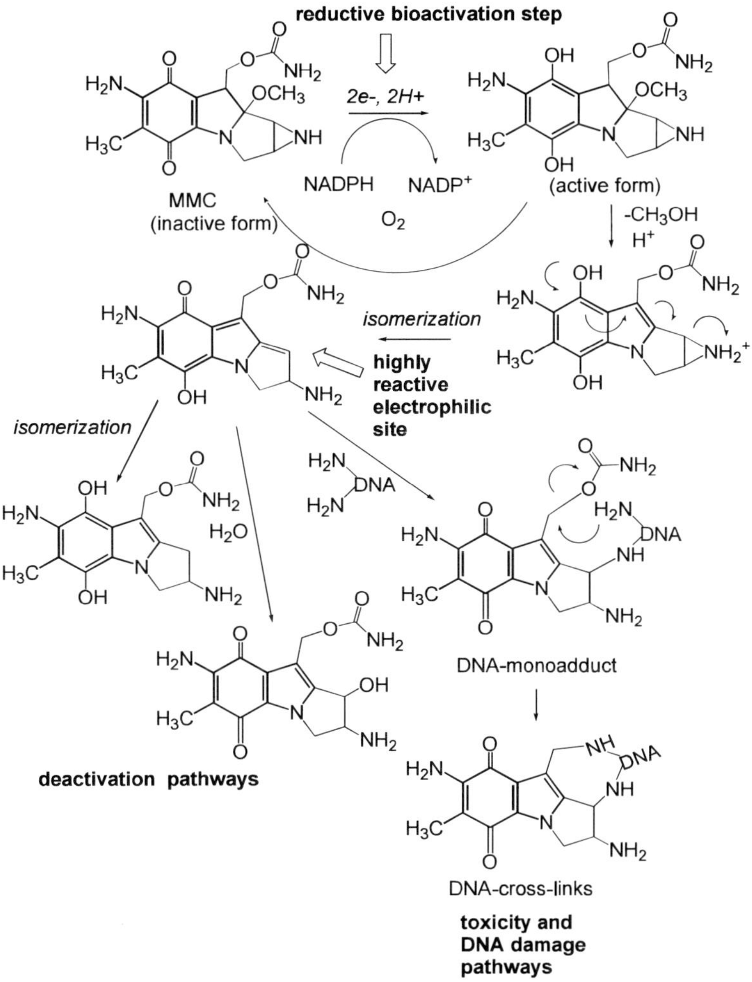

Figure 1. Schematic illustration of MMC bioactivation and DNA damage pathways.

Diepoxybutane

Diepoxybutane has long been known as a directly acting mutagen following 1,3-butadiene biotransformation (Olszewska and Kilbe 1975). Subsequent investigations elucidated the role of GSH and GSH-related activities (GSH S-transferase, GST) in coping with DEB-induced toxicity, as shown in Figure 2.

Pelin et al. (1996) reported that GSTT1 activity was involved in a major detoxification pathway for DEB, and Landi et al. (1996) found that DEB-induced sister chromatid exchange (SCE) was dependent on the expression of GSTT1 and GSTM1, while it was enhanced in GSTT1-null homozygotes; moreover, GSH depletion was found to enhance DEB toxicity (Vlachodimitropoulos et al. 1997; Spanò et al. 1998) as summarized in Table 2. Other studies reported on redox modulation of DEB toxicity, such as expression of thioredoxin (Trx), GST and GSH peroxidase (GSH-Px) in FA cells, DEB-induced ROS formation, and inhibition of CYP2E1 activity (Ruppitsch et al. 1998; Erexson and Tindall 2000; Davies et al. 2005;

Yadavilli et al. 2007; Hartman et al. 2014). The toxicity mechanisms displayed by DEB altogether suggest that its activity is due to its potent cross-linking effects and alkylating properties. These stem from the peculiar 1,3-diepoxidation pattern of DEB (Vlachodimitropoulos et al. 1997) in which the ring tension accounts for the marked reactivity toward nucleophiles, and the double epoxide ring allows for the formation of interstrand crosslinks with many DNA sequences (Figure 2). A substantial body of evidence suggests that individual sensitivity to the genotoxic and cytotoxic effects of DEB depend on oxygen levels and result in a dose-related GSH depletion (Boogaard and Bond 1996). Two glutathione-related activities, glutathione-S-transferase (GST), and GSH peroxidase (GPx) have been implicated in the detoxification of DEB-induced DNA damage (Erexon and Tindall 2000; Spanò et al. 1998). In addition to conjugation with GSH, a possible transformation route of DEB is hydrolysis (Boogaard and Bond 1996). A comparative screening of the microsomal hydrolysis of DEB using tissues of rats, mice, and humans was found to give two main metabolites - erythritol and anhydroerythritol. It is tempting to speculate that diol and polyol products of DEB hydrolysis, though expectedly inactive *per se*, may be sources of cytotoxic aldehydes (e.g., glyoxal) following oxidation and retroaldol reactions under OS conditions (Boogaard and Bond 1996). Both DEB and MMC exposures were found to induce concentration-related DNA oxidative damage (8-hydroxy-deoxyguanosine, 8-OHdG) in sea urchin embryos (Pagano et al. 2001).

Nitrogen Mustards: Melphalan and Cyclophosphamide

The alkylating action of several nitrogen mustards has long been related to GSH depletion that is involved in detoxification and repair of cellular injury, along with induction of oxidative damage and with increased hypoxia sensitivity (Robson et al. 1987; Rappeneau et al. 1999; Ucar et al. 2007; Zheng et al. 2013) (Table 3). In the case of MEL, toxicity is enhanced by GSH depletion and hypoxia (Arrick and Nathan 1984; Roizin-Towle 1985; Lunel-Orsini et al. 1995), by increasing DNA fragmentation, peroxide accumulation and endogenous formation of H_2O_2 (Vahrmeijer et al. 1999; de Wilt et al. 1999; Troyano et al. 2001; Matsura et al. 2004). Altogether, the direct MEL-associated crosslinking activity can be related to the multi-fold relationships of MEL toxicity with GSH levels, GSH-related activities, and with MEL-induced prooxidant state.

Analogous to MEL, as shown in Table 3, CP toxicity was found to relate with GSH depletion, and CP biotransformation resulted in ROS-forming active metabolites, along with oxidative DNA damage and showing protection by antioxidants (Crook et al. 1986; LeBlanc and Waxman 1990; Murata et al. 2004; Selvakumar et al. 2006; Sadir et al. 2007; Liu et al. 2012) (Figure 3).

A specific relationship of CP toxicity with FA phenotype was reported by Joenje and Oostra (1986), who found a direct clastogenic effect of CP in FA lymphocyte cultures, suggesting a spontaneous propensity of FA cells in CP bioactivation. In the lack of subsequent studies confirming this relevant observation, one may envisage pharmacologic investigations with up-to-date methods and tools, evaluating FA cell proficiency in bioactivating CP and MMC.

Table 1. Historical summary of early reports of MMC bio-reductive activation (1960's-1980's)*

Authors (years)	Main Findings
Iyer & Szybalski (1963;1964)	MMC *in vitro* effect on DNA requires a cell extract addition; formation of links between complementary DNA strands by metabolically activated MMC quinone-
Handa & Sato (1975)	containing anticancer chemicals form semiquinone radicals
Bachur et al. (1979)	NADPH cytochrome P-450 reductase catalyzes the single-electron reduction of quinone antibiotics (including MMC); MMC is activated to free radical intermediates which transfer their single electron to O_2 to form $^{\bullet}O_2^-$
Kennedy et al. (1980)	MMC toxicity is enhanced in hypoxic cells and requires NADPH-generating system;
Teicher et al. (1981) Tomasz & Lipman (1981) Rockwell et al. (1982)	ROS formation during MMC bioactivation
Trush et al. (1982)	MMC biotransformation generates lipid peroxidation; lipid peroxidation was inhibited by SOD, GSH, ascorbic acid, and CAT
Gutteridge et al. (1984)	MMC stimulates deoxyribose degradation with peroxidation products under hypoxia. This damage is inhibited by $^{\bullet}OH$ radical scavengers, Fe chelators, CAT and SOD

* Abbreviations: MMC = mitomycin C; ROS = reactive oxygen species; SOD = superoxide dismutase; CAT = catalase.

Table 2. Historical summary of published evidence for DEB toxicity mechanisms*

Authors (years)	Main Findings
Pelin et al. (1996)	GSTT1 activity as a major detoxification pathway for DEB
Landi et al. (1996)	SCE induction by DEB is enhanced in GSTT1-null homozygotes
Spanò et al. (1998)	increased DEB sensitivity in lymphocytes from GSTT-deficient donors
Kennedy et al. (1980)	glutathione depletion enhances DEB toxicity
Ruppitsch et al. (1998)	overexpression of thioredoxin in FA cells abolished MMC- and DEB-induced DNA damage
Erexson & Tindall (2000)	GSH-Px is involved in the detoxification of DEB-induced DNA damage; DEB causes ROS formation
Davies et al. (2005)	DEB sensitivity was significantly increased in GSTT1-null FA patients; GSTM1 influenced time to BMF in FA-C patients
Yadavilli et al. (2007)	DEB induced apoptosis and ROS formation; trans-membrane potential in DEB-induced apoptosis
Hartman et al. 2014	DEB inhibits CYP2E1 activity

Abbreviations: DEB = diepoxybutane; GSTT = GSH S-transferase theta; SCE= sister chromatid exchange; CYP = cytochrome P 450.

Defective DNA Repair: The Shared Axiom Revisited

By far the most shared opinion, defective DNA repair with excess sensitivity to crosslinkers is associated with FA definition and provides the basis for most of the FA literature up to recent reviews (Kim and D'Andrea 2012; Kupfer 2013).

The pioneering studies by Sasaki (1975) tested the susceptibility to chromosome breakage of peripheral blood lymphocytes from FA patients exposed to a number of clastogens (including MMC). The increased susceptibility to compounds resulting in DNA ICL was interpreted as an indication that the FA cells are defective in DNA repair mechanisms. No mention was provided to the toxicity mechanisms of the tested clastogens, namely related to MMC requirement of bioactivation, as available in the previous and contemporary literature (e.g., Iyer and Szybalski 1964; Handa and Sato 1975).

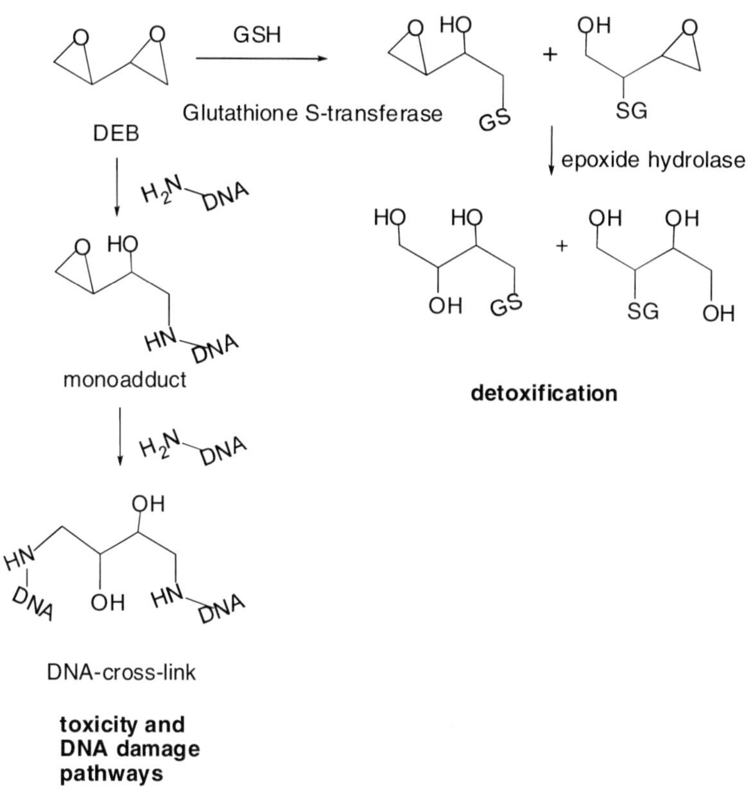

Figure 2. Schematic view of DEB-induced DNA damage pathways and mechanism of detoxification by glutathione (GSH).

Fujiwara investigated excision repair ability in FA cells in terms of MMC cell killing and by DNA sucrose sedimentation, compared to HeLa cells and Xeroderma pigmentosum (XP) cells (Fujiwara 1982). The results showed that FA cell DNA underwent faster sedimentation after MMC treatment, «suggesting a possible impairment in FA cell's ability to unhook crosslinks».

Table 3. Historical summary of published evidence for toxicity mechanisms of melphalan and cyclophosphamide*

Authors (years)	Main Findings
MELPHALAN	
Roizin-Towle (1985)	increased toxicity by GSH depletion and hypoxia
Lunel-Orsini et al. (1995)	GSH depletion increased MEL toxicity
Vahrmeijer et al. (1999)	GSH depletion caused a 4-fold increase in DNA fragmentation and increase in apoptotic cells
de Wilt et al. (1999)	hypoxia enhanced anti-tumour activity of MEL
Troyano et al. (2001)	GSH depletion enhanced peroxide accumulation
Matsura et al. (2004)	H_2O_2 formation, phosphatidylserine oxidation, cyt C release in cytosol
CYCLOPHOSPHAMIDE	
Crook et al. (1986)	CP metabolite acrolein effective in GSH depletion
LeBlanc & Waxman (1990)	CP/acrolein modulate hepatic P-450 expression
Stankiewicz & Skrzydlewska (2003)	CP is biotransformed into ROS-forming active metabolites; decreased SOD, GPx, GR and CAT; decreased GSH, Vit C and E
Murata et al. (2004)	4-hydroperoxyCP caused oxidative DNA damage
Selvakumar et al. (2006)	ALA protects of CP-induced DNA damage
Sadir et al. (2007)	protection by antioxidants of CYP-induced OS
Kasapović et al. (2010)	CP causes oxidative DNA damage and decreases antioxidant activities (SOD, GPx, GR) and GSH
Liu et al. (2012)	CP/acrolein impairs cytoskeleton and induces OS

* Abbreviations: GPx = glutathione peroxidase; GR = glutathione reductase; 8-oxodG = 8-Oxo-2'-deoxyguanosine; ALA = α-lipoic acid.

Also this author failed to verify - or to cite from the available literature - whether the excess MMC sensitivity of FA cells might be due to an abnormal propensity to activate MMC to its directly acting metabolites (Fujiwara 1982).

Latt et al. (1975) found a decreased sister chromatid exchange (SCE) induction in FA cells *vs.* normal cells, while MMC treatment resulted in partial mitosis suppression and an increase in chromatid breaks and rearrangements. The authors suggested that chromosomal breaks and rearrangements in FA cells might result from a defect in repair of DNA damage.

Unlike the above studies, Auerbach and Wolman (1976) investigated the effects of DEB and methyl methane sulfonate (MMS) as directly acting mutagens, by prolonged exposure of FA fibroblasts *vs.* XP and *vs.* control cells, and by evaluating the frequency of chromosomal aberrations after DEB or MMS removal. This study was designed by assuming the direct mutagenic action for DEB, as still currently recognized, while the GSH-related modulation of DEB toxicity (as shown in Table 2) would be reported some twenty years later (Pelin et al. 1996; Vlachodimitropoulos et al. 1997).

Raj and Heddle (1980) tested the induction of chromosomal breakage in FA and normal cells with or without MMC treatment and cultured with SOD, or CAT, or l-cysteine, reaching a quite singular conclusion. These authors found that antioxidant treatments resulted in decreased chromosomal breaks both in FA and in normal cells, thus they stated that the effect was not specific, by hence concluding for the occurrence of DNA repair deficiency. Opposite conclusions were provided by Nagasawa and Little (1983), who found that survival of MMC-treated FA cells was increased by adding SOD, unlike MMC-treated normal cells. These authors suggested that the hypersensitivity to «MMC cytoxicity in FA cells may involve among other factors a deficiency in the inactivation of certain free radical intermediates» (Nagasawa and Little 1983).

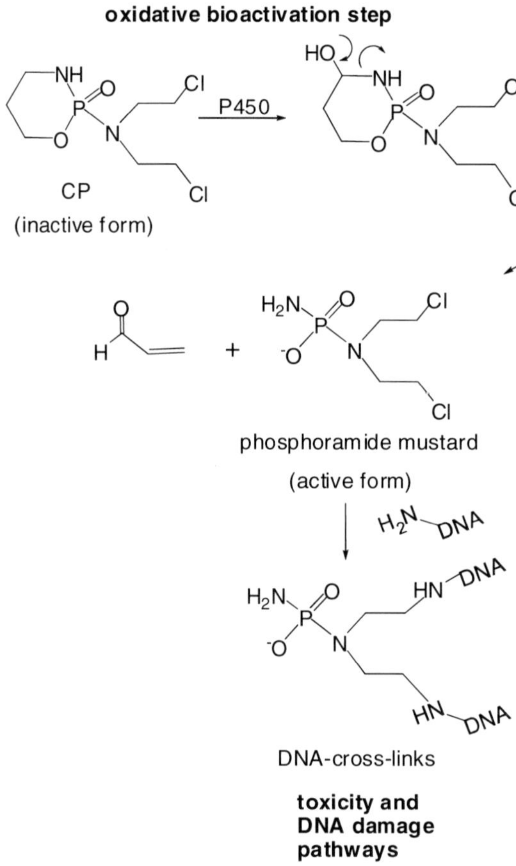

Figure 3. Schematic illustration of CP bioactivation and toxicity pathways.

A major support to the theory of DNA repair deficiency was provided by Ishida and Buchwald (1982). This report evaluated growth inhibition of FA lymphoblasts exposed to a number of mono- and bifunctional crosslinking agents including, among others, MMC and MEL, and found that FA cells are more sensitive than normal cells to all DNA crosslinking agents examined. Interestingly, these authors stated that «MMC requires activation by a NADPH-generating system (Iyer and Szybalski 1963) before it will cross-link DNA while melphalan does not (Parsons et al. 1981), yet FA cells have similar sensitivities to the two drugs. It is probable, therefore, that the defect(s) in FA is at the level of repair of the damage in DNA» (Ishida and Buchwald 1982). About MMC biotransformation, the authors provided a quite outdated citation from Iyer and Szybalski (1963), published almost twenty years before this report, by disregarding the recent and contemporary literature providing evidence for redox-related MMC biotransformation (e.g., Handa and Sato 1975; Bachur et al. 1979; Kennedy et al. 1980). On the other hand, one should recognize that studies of redox-related toxicity mechanisms of MEL were not yet available. Thus, this basic report on FA-associated DNA repair deficiency reached a conclusion partly unaware of the up-to-date literature (MMC bioactivation), and partly surpassed by following investigations (MEL-associated redox toxicity mechanisms).

Altogether, the founding publications supporting the theory of DNA repair deficiency in FA either failed to consider the available literature pointing to MMC bioactivation requirements, or could not foresee the subsequent evidence for redox toxicity (GSH-related) mechanisms of DEB and MEL. That theory, though presently shared almost universally in the FA research community, cannot be regarded as an undisputable truth, as far as its origins can be critically reconstructed in a historical framework, displaying some inborn pitfalls as discussed here.

From Pharmacology to Research Prospects

Unconfined to a doctrinal issue, recognizing the basic defect(s) in FA pathogenesis involves the design of FA patients' clinical management. The prevailing opinion attributing FA's basic defect to impaired DNA repair has led to the current state-of-art in clinical practice, as yet confined to androgen therapy and/or to optimized transplantation protocols.

On the other hand, a reappraisal of crosslinker toxicity mechanisms in FA cells may lead to reconsider the dysmetabolic condition(s) associated to FA phenotype, mainly related to the redox anomalies detected in a body of evidence based on molecular, *in vitro* and *in vivo* studies.

Thus, elucidating the toxicity mechanisms of MMC and DEB may turn to revolutionary consequences both of theoretical nature and with clinical relevance in FA patients' management.

References

Arrick, B. A. & Nathan, C. F. (1984). Glutathione metabolism as a determinant of therapeutic efficacy: a review. *Cancer Res.*, *44*, 4224-4232.

Auerbach, A. D. & Wolman, S. R. (1976). Susceptibility of Fanconi's anaemia fibroblasts to chromosome damage by carcinogens. *Nature*, *261*, 494-496.

Auerbach, A. D., Warburton, D., Bloom, A. D. & Chaganti, R. S. (1979). Preliminary communication: prenatal detection of the Fanconi Anemia gene by cytogenetic methods. *Am. J. Hum. Genet.*, *31*, 77-81.

Bachur, N. R., Gordon, S. L., Gee, M. V. & Kon, H. (1979). NADPH cytochrome P-450 reductase activation of quinone anticancer agents to free radicals. *Proc. Natl. Acad. Sci. USA*, *76*, 954-957.

Boogaard, P. J. & Bond, J. A. (1996). The role of hydrolysis in the detoxification of 1,2:3,4-diepoxybutane by human, rat, and mouse liver and lung in vitro. *Toxicol. Appl. Pharmacol.*, *141*, 617-627.

Crook, T. R., Souhami, R. L., Whyman, G. D. & McLean, A. E. (1986). Glutathione depletion as a determinant of sensitivity of human leukemia cells to cyclophosphamide. *Cancer Res.*, *46*, 5035-5038.

Davies, S. M., Radloff, G. A., DeFor, T. E., Levran, O., Batish, S. D., Hanenberg, H. & Auerbach, A. D. (2005). GST genotype may modify clinical phenotype in patients with Fanconi anaemia. *Br. J. Haematol.*, *131*, 118-122.

de Wilt, J. H. W., Manusama, E. R., van Tiel, S. T., van Ijken, M. G., ten Hagen, T. L. & Eggermont, A. M. (1999). Prerequisites for effective isolated limb perfusion using tumour necrosis factor alpha and melphalan in rats. *Brit. J. Cancer*, *80*, 161–166.

Erexson, G. L. & Tindall, K. R. (2000). Reduction of diepoxybutane-induced sister chromatid exchanges by glutathione peroxidase and erythrocytes in transgenic Big Blue mouse and rat fibroblasts. *Mutat.Res.*, *447*, 267-274.

Fujiwara, Y. & Tatzumi, M. (1975). Repair of mitomycin C damage to DNA in mammalian cells and its impairment in Fanconi's anemia cells. *Biochem. Biophys. Res. Commun.*, *66*, 592-598.

Fujiwara, Y. (1982). Defective repair of mitomycin C crosslinks in Fanconi's anemia and loss in confluent normal human and xeroderma pigmentosum cells. *Biochim. Biophys. Acta* *699*, 217-225.

Gutteridge, J. M., Quinlan, G. J. & Wilkins, S. (1984). Mitomycin C-induced deoxyribose degradation inhibited by superoxide dismutase. A reaction involving iron, hydroxyl and semiquinone radicals. *FEBS Lett.*, *167*, 37-41.

Handa, K. & Sato, S. (1975). Generation of free radicals of quinone group-containing anti-cancer chemicals in NADPH-microsome system as evidenced by initiation of sulfite oxidation. *Gann*, *66*, 43-47.

Hartman, J. H., Miller, G. P. & Boysen, G. (2014). Inhibitory potency of 4-carbon alkanes and alkenes toward CYP2E1 activity. *Toxicology*, *318*, 51-58.

Ishida, R. & Buchwald, M. (1982). Susceptibility of Fanconi's anemia lymphoblasts to DNA-cross-linking and alkylating agents. *Cancer Res.*, *42*, 4000-4006.

Iyer, V. N. & Szybalski, W. (1963). A molecular mechanism of mitomycin action: Linking of complementary DNA strands. *Proc. Natl. Acad. Sci. USA*, *50*, 355-362.

Iyer, V. N. & Szybalski, W. (1964). Mitomycins and porfiromycin: Chemical mechanism of activation and cross-linking of DNA. *Science*, *145*, 55-58.

Joenje, H. & Oostra, A. B. (1986). Clastogenicity of cyclophosphamide in Fanconi's anemia lymphocytes without exogenous metabolic activation. *Cancer Genet.Cytogenet.*, *22*, 339-345.

Kasapović, J., Pejić, S., Stojiljković, V., Todorović, A., Radošević-Jelić, L., Saičić, Z. S. & Pajović, S. B. (2010). Antioxidant status and lipid peroxidation in the blood of breast cancer patients of different ages after chemotherapy with 5-fluorouracil, doxorubicin and cyclophosphamide. *Clin. Biochem.*, *43*, 1287-1293.

Kennedy, K. A., Rockwell, S. & Sartorelli, A. C. (1980). Preferential activation of mitomycin C to cytotoxic metabolites by hypoxic tumor cells. *Cancer Res.*, *40*, 2356-2360.

Kim, H. & D'Andrea, A. D. (2012). Regulation of DNA cross-link repair by the Fanconi anemia/BRCA pathway. *Genes Dev.*, *26*, 1393-1408.

Kupfer, G. M. (2013). Fanconi anemia: a signal transduction and DNA repair pathway. *Yale J. Biol. Med.*, *86*, 491-497.

Landi, S., Ponzanelli, I., Hirvonen, A., Norppa, H., & Barale R. (1996). Repeated analysis of sister chromatid exchange induction by diepoxybutane in cultured human lymphocytes: effect of glutathione S-transferase T1 and M1 genotype. *Mutat. Res., 351*, 79-85.

Latt, S. A., Stetten, G., Juergens, L. A., Buchanan, G. R. & Gerald, P. S. (1975). Induction by alkylating agents of sister chromatid exchanges and chromatid breaks in Fanconi's anemia. *Proc. Natl. Acad. Sci. USA*, *72*, 4066-4070.

LeBlanc, G. A. & Waxman, D. J. (1990). Mechanisms of cyclophosphamide action on hepatic P-450 expression. *Cancer Res., 50*, 5720-5726.

Liu, F., Li, X. L., Lin, T., He, D. W., Wei, G. H., Liu, J. H. & Li, L. S. (2012). The cyclophosphamide metabolite, acrolein, induces cytoskeletal changes and oxidative stress in Sertoli cells. *Mol. Biol. Rep., 39*, 493-500.

Lunel-Orsini, C., Buttin, G., De Sain & Vincent, B. R. (1995). A glutathione depletion selectively imposed on mu glutathione S-transferase overproducing cells increases nitrogen mustard toxicity. *Biochem. Pharmacol., 49*, 329-338.

Matsura, T., Kai, M., Jiang, J., Babu, H., Kini, V., Kusumoto, C., Yamada, K. & Kagan, V. E. (2004). Endogenously generated hydrogen peroxide is required for execution of melphalan-induced apoptosis as well as oxidation and externalization of phosphatidylserine. *Chem. Res. Toxicol., 17*, 685-696.

Murata, M., Suzuki, T., Midorikawa, K., Oikawa, S. & Kawanishi, S. (2004). Oxidative DNA damage induced by a hydroperoxide derivative of cyclophosphamide. *Free Radic. Biol. Med., 37*, 793-802.

Nagasawa, H. & Little, J. B. (1983).Suppression of cytotoxic effect of mitomycin-C by superoxide dismutase in Fanconi's anemia and dyskeratosis congenita fibroblasts. *Carcinogenesis, 4*, 795-799.

Norppa, H. & Sorsa, M. (1993). Genetic toxicity of 1,3-butadiene and styrene. *IARC Sci. Publ., 127*, 185-193.

Olszewska, E. & Kilbe, B. J. (1975). The mutagenic activity of diepoxybutane in yeast. *Mutat. Res., 33*, 383-390.

Pagano, G., Degan, P., De Biase, A., Iaccarino, M. & Warnau, M. (2001). Diepoxybutane and mitomycin C toxicity is associated with the induction of oxidative DNA damage in sea urchin embryos. *Hum. Exp. Toxicol., 20*, 651-655.

Parsons, P. G., Carter, F. B., Morrison, L. & Regius, M. S. (1981). Mechanism of melphalan resistance developed in vitro in human melanoma cells. *Cancer Res., 41*, 1525-1534.

Pelin, K., Hirvonen, A. & Norppa, H. (1996). Influence of erythrocyte glutathione S-transferase T1 on sister chromatid exchanges induced by diepoxybutane in cultured human lymphocytes. *Mutagenesis, 11*, 213-215.

Raj, A. S. & Heddle, J. A. (1980).The effect of superoxide dismutase, catalase and L-cysteine on spontaneous and on mitomycin C induced chromosomal breakage in Fanconi's anemia and normal fibroblasts as measured by the micronucleus method. *Mutat. Res., 78*, 59-66.

Rappeneau, S., Baeza-Squiban, A., Braut-Boucher, F., Aubery, M., Gendron, M. C. & Marano, F. (1999). Use of fluorescent probes to assess the early sulfhydryl depletion and oxidative stress induced by mechlorethamine in human bronchial epithelial cells. *Toxicol.* In vitro, *13*, 765-771.

Robson, C. N., Lewis, A. D., Wolf, C. R., Hayes, J. D., Hall, A., Proctor, S. J., Harris, A. L. & Hickson, I. D. (1987). Reduced levels of drug-induced DNA cross-linking in nitrogen mustard-resistant Chinese hamster ovary cells expressing elevated glutathione S-transferase activity. *Cancer Res.*, *47*, 6022-6027.

Rockwell, S., Kennedy, K. A. & Sartorelli, A. C. (1982). Mitomycin-C as a prototype bioreductive alkylating agent: in vitro studies of metabolism and cytotoxicity. *Int. J. Radiat. Oncol. Biol. Phys.*, *8*, 753-755.

Roizin-Towle, L. (1985). Selective enhancement of hypoxic cell killing by melphalan via thiol depletion: in vitro studies with hypoxic cell sensitizers and buthionine sulfoximine. *J. Natl. Cancer Inst.*, *74*, 151-157.

Ruppitsch, W., Meisslitzer, C., Hirsch-Kauffmann, M. & Schweiger, M. (1998). Overexpression of thioredoxin in Fanconi anemia fibroblasts prevents the cytotoxic and DNA damaging effect of mitomycin C and diepoxybutane. *FEBS Lett.*, *422*, 99-102.

Sadir, S., Deveci, S., Korkmaz, A. & Oter, S. (2007). Alpha-tocopherol, beta-carotene and melatonin administration protects cyclophosphamide-induced oxidative damage to bladder tissue in rats. *Cell Biochem. Funct.*, *25*, 521-526.

Sasaki, M. S. & Tonomura, A. (1973). A high susceptibility of Fanconi's anemia to chromosome breakage by DNA cross-linking agents. *Cancer Res.*, *33*, 1829-1836.

Sasaki, M. S. (1975). Is Fanconi's anaemia defective in a process essential to the repair of DNA cross links? *Nature*, *257*, 501-503.

Schroeder, T. M., Anschütz, F. & Knopp, A. (1964).Spontaneous chromosome aberrations in familial panmyelopathy. *Humangenetik*, *1*, 194-196.

Selvakumar, E., Prahalathan, C., Sudharsan, P. T. & Varalakshmi, P. (2006).Chemoprotective effect of lipoic acid against cyclophosphamide-induced changes in the rat sperm. *Toxicology*, *217*, 71-78.

Spanò, M., Cordelli, E., Leter, G. & Pacchierotti, F. (1998). Diepoxybutane cytotoxicity on mouse germ cells is enhanced by in vivo glutathione depletion: a flow cytometric approach. *Mutat. Res.*, *397*, 37-43.

Stankiewicz, A. & Skrzydlewska, E. (2003). Protection against cyclophosphamide-induced renal oxidative stress by amifostine: the role of antioxidative mechanisms. *Toxicol. Mech. Methods*, *13*, 301-308.

Strese, S., Fryknäs, M., Larsson, R. & Gullbo, J. (2013). Effects of hypoxia on human cancer cell line chemosensitivity. *BMC Cancer*, *13*, 331.

Teicher, B. A., Lazo, J. S. & Sartorelli, A. C. (1981). Classification of antineoplastic agents by their selective toxicities toward oxygenated and hypoxic tumor cells. *Cancer Res.*, *41*, 73-81.

Troyano, A., Fernández, C., Sancho, P., de Blas, E. & Aller, P. (2001). Effect of glutathione depletion on antitumor drug toxicity (apoptosis and necrosis) in U-937 human promonocytic cells. The role of intracellular oxidation. *J. Biol. Chem.*, *276*, 47107-47115.

Trush, M. A., Mimnaugh, E. G., Ginsburg, E. & Gram, T. E. (1982). Studies on the in vitro interaction of mitomycin C, nitrofurantoin and paraquat with pulmonary microsomes.Stimulation of reactive oxygen-dependent lipid peroxidation. *Biochem. Pharmacol.*, *31*, 805-814.

Ucar, M., Korkmaz, A., Reiter, R. J., Yaren, H., Oter, S., Kurt, B. & Topal, T. (2007). Melatonin alleviates lung damage induced by the chemical warfare agent nitrogen mustard. *Toxicol. Lett.*, *173*, 124-131.

Vahrmeijer, A. L., van Dierendonck, J. H., Schutrups, J., van de Velde, C. J. & Mulder, G. J. (1999). Effect of glutathione depletion on inhibition of cell cycle progression and induction of apoptosis by melphalan (L-phenylalanine mustard) in human colorectal cancer cells. *Biochem. Pharmacol.*, *58*, 655-664.

Vlachodimitropoulos, D., Norppa, H., Autio, K., Catalán, J., Hirvonen, A., Tasa, G., Uusküla, M., Demopoulos, N. A. & Sorsa, M. (1997). GSTT1-dependent induction of centromere-negative and -positive micronuclei by 1,2:3,4-diepoxybutane in cultured human lymphocytes. *Mutagenesis*, *12*, 397-403.

Yadavilli, S., Martinez-Ceballos, E., Snowden-Aikens, J., Hurst, A., Joseph, T., Albrecht, T. & Muganda, P. M. (2007). Diepoxybutane activates the mitochondrial apoptotic pathway and mediates apoptosis in human lymphoblasts through oxidative stress. *Toxicol.* In vitro, *21*, 1429-1441.

Zheng, R., Po, I., Mishin, V., Black, A. T., Heck, D. E., Laskin, D. L., Sinko, P. J., Gerecke, D. R., Gordon, M. K. & Laskin, J. D. (2013). The generation of 4-hydroxynonenal, an electrophilic lipid peroxidation end product, in rabbit cornea organ cultures treated with UVB light and nitrogen mustard. *Toxicol. Appl. Pharmacol.*, *272*, 345-355.

In: Fanconi Anemia and Oxidative Stress
Editor: Giovanni Pagano
ISBN: 978-1-63482-297-8

Chapter 4

Involvement of Mitochondria in the Pathology of Fanconi Anemia

Pavithra Shyamsunder*[1]*, Rama S. Verma*[1]*,
***Sudit S. Mukhopadhyay*[2] *and Alex Lyakhovich*[*,3,4]**
[1]Stem cell and Molecular biology Laboratory, Department of Biotechnology, Indian Institute of Technology Madras, Chennai, India
[2]Department of Biotechnology, National Institute of Technology, Durgapur, West Bengal, India
[3]Novosibirsk Institute of Molecular Biology and Biophysics, Russia
[4]Queen's University Belfast, School of Biological Sciences Medical Biology Centre, Belfast, UK

Abstract

Although major pathological conditions in FA are outcomes of the defects in DNA damage and repair machinery, recent findings suggest that bioenergetic pathways are compromised as well. A few FA proteins have functions that include redox balance, apoptosis, and energy metabolism; and recent data have established respiratory defects in FA cells, proposing that mitochondrial dysfunction (MDF) is involved in FA. This data was further supported by the finding that FA cells have elevated levels of mitochondrial ROS, decreased Mitochondrial Membrane Potential, decreased ATP production, impaired oxygen uptake and changes in the morphology of mitochondria. This was also accompanied by inactivation of enzymes that are essential in the energy production pathway. These findings suggest that mitochondrial activity and its metabolism may be a point of convergence in the pathological manifestations of FA. Mitochondrial dysfunction may thus be considered as an important phenomenon in FA which might account for most of the findings observed in FA's clinical phenotype. Exploring MDF in FA will provide new facets to the development of diagnostic and therapeutic strategies for FA.

Keywords: Fanconi Anemia, mitochondria, oxidative stress, ROS, NOS

---

[*]Corresponding author: E-mail: lyakhovich@gmail.com.

Introduction

Most scientific research on FA points to a deficiency in DNA repair in response to DNA damage. The damage can be endogenous or induced by exogenous agents such as mitomycin C (MMC), diepoxybutane (DEB), cis-platin, cyclophosphamide, hexavalent chromium compounds and 8-methoxypsoralen plus near ultraviolet light (PUVA) (Auerbach and Wolman 1976; Alter 1993, 1996; Ishida and Buchwald 1982; Poll et al. 1982; Moustacchi and Diatloff-Zito 1985; Wang and D'Andrea 2004; Taniguchi and D'Andrea 2006). However, the sensitivities of FA cells to MMC and DEB have also been attributed to interference with the redox cycle and point to a role of oxidative stress (OS) in the onset of the disease (Korkina et al. 1992; Pagano and Youssoufian 2003; Pagano et al. 2005). The relevance of OS in FA was initially suggested by Nordenson (1977) and by Joenje et al. (Joenje et al. 1979; 1981) and then in other early studies (Schindler and Hoehn 1988; Pagano et al. 1997) led to a body of evidence pointing to OS as a major phenotypic hallmark in FA. This idea is further strengthened by the fact that FA proteins have functions that extend into regulation of redox states and energy metabolism (Cumming et al. 2001). Interestingly, a number of FA patients reveal hydrocephalia, ventriculoperitoneal shunts, erythrophagocytosis and type II diabetes mellitus - common hallmarks of diseases that involve impaired oxygen metabolism and mitochondrial damage (Pagano et al. 2012). FA cells are sensitive to oxidants (Saadatzadeh et al. 2004; Rani et al. 2008) , and accumulation of oxidized proteins has been characterized for several FA subgroups (Lyakhovich and Surrallés 2010). Clinical phenotypes of FA suggest altered functions of mitochondria, organelles responsible for cellular energetic metabolism, and also serving as an important producer and the most susceptible target from reactive oxygen species (ROS). Therefore, alterations in FA mitochondria may be a clue to the susceptibility of FA cells to oxidative stimuli.

Mitochondria's Role in Oxidative Stress

Oxidative stress is generally defined as an imbalance that favors the production of ROS over antioxidant defenses (Poljsak et al. 2013). However, the precise mechanisms by which ROS produce cellular injury remain elusive. In eukaryotic systems, mitochondria are associated with the process of energetic metabolism (Fariss et al. 2001) and are the principal source of ROS formation (Turrens 2003). Due to such proximity of ROS generating cascades, mitochondria are also the prime target for ROS. Increasing levels of ROS have been implicated in evading cellular antioxidant defense mechanisms and thus causing mitochondrial damage. This damage is augmented when the endogenous antioxidant network and repair systems are overwhelmed (Zhang et al. 2001) as observed in FA. ROS are implicated in a wide variety of pathologies, including malignant diseases, type II diabetes, atherosclerosis, chronic inflammatory processes, ischemia/reperfusion injury, and several neurodegenerative diseases (Fariss and Zhang 2003; Pagano et al. 2014). At the same time, ROS play a regulatory role in cellular metabolic processes by activation of various enzymatic cascades as well as several transcription factors (Gupta et al. 1999; Brown et al. 2004). The majority of ROS are products of mitochondrial respiration (Murphy 2009). Approximately 1%–3% of the molecular oxygen consumed during normal physiological respiration is converted into superoxide radicals. Although ROS targets vary from cell to cell types, the impact of oxidative damage over the cell compartment is different. In particular, mitochondrial DNA (mtDNA) that encodes 13 polypeptides, 22 transfer RNAs (tRNAs), and 2 ribosomal RNAs (rRNAs) are an obvious target for the oxidative damage. The mtDNA is extensively affected by ROS so are the protein products of both the mitochondrial and nuclear genomes engaged into functional respiratory complexes (Fariss et al. 2005). A potentially deleterious effect of ROS production in mitochondria is facilitation of Ca^{2+}-dependent mitochondrial permeability transition (MPT), which plays a key role in certain modes of cell death. Hence, both OS and impaired Ca^{2+} homeostasis contribute to mitochondria-mediated cellular damage (Duchen 2000). MPT is a mechanism causing mitochondrial failure, which can lead to necrosis owing to ATP depletion or to caspase-dependent apoptosis if MPT induction occurs in a subpopulation of mitochondria and remaining organelles are still able to maintain the MMP and produce ATP (Orrenius et al. 2003).

Redox-Related Functions of FA Proteins

There have been reports of interactions of the FA proteins and involvement of the FA proteins in redox-related functions. The FANCC protein has been found to be associated with NADPH cytochrome P450 reductase and involved in redox related activities (Kruyt et al. 1998). The involvement of FANCC with GSTP1 has also been well studied, wherein FANCC was involved in prevention of protein oxidation and thus revealed a novel mechanism of enzyme regulation during apoptosis (Cumming et al. 2001). A protein-protein interaction study identified over 70 direct interacting partners of FANCC, FANCA and FANCG. These included proteins involved in the processes of transcription, oxidative metabolism and transport mechanisms (Reuter et al. 2003) and multiple genome wide expression study of FANCC lymphoblasts had lead to the identification of genes involved in the process of inflammation (Zanier et al. 2004) and intracellular signaling (Shyamsunder et al. 2013a). For example, an expression profile of Indian FA patients revealed deregulation of genes involved in the process of transcription, intracellular transport and metabolic processes (Shyamsunder et al. 2013a). The interaction of the FANCG protein with a cytochrome P450 2E1 (CYP2E1) was identified (Futaki et al. 2002) and it is known to be involved in redox biotransformation of xenobiotics of compounds such as MMC (Szybalski and Iyer 1964; Pritsos and Sartorelli 1986; Dusre et al. 1990; Clarke et al. 1997; Penketh et al. 2001). The influence of redox states on FANCA and FANCG has been demonstrated by their ability to form disulphide bonds in the FA complex (Park et al. 2004). Recently, the occurrence of mitochondrial defects was demonstrated by Mukhopadhyay et al. (2006) and by Kumari et al. (2014).

Studies using Fanca -/- Trp53 -/- double knockout mice revealed that deficiency in the Fanca gene stimulated a DNA damage response to oxidative DNA and a p53-dependent growth arrest. It was demonstrated that OS induced p53 response in Fanca -/- cells was likely due to accumulation of unrepaired DNA damage (Rani et al. 2008). There are also reports that suggest that the activation of p53 leads to an increase in ROS levels that leads to cell death by probably interfering with mitochondrial function and/or integrity and that the ROS levels in turn through a feedback mechanism aid in stabilizing p53 and hence contributing to its heightened activity (Fridman and Lowe 2003).

Mitochondrial Defects Observed in FA Cells and the Role of Cross-Linking Agents

FA cells display an increased activity of the antioxidant phospholipid–hydroperoxide–gluthatione-peroxidase (Ruppitsch et al. 1997). An pro-oxidant state accompanied by increase in ROS and interference with cellular redox state (Degan et al. 1995) and ATP production (Bogliolo et al. 2002), sensitivity to oxidant stimuli (Saadatzadeh et al. 2004, Rani et al. 2008) and accumulation of oxidized proteins (Lyakhovich and Surrallés 2010) have been characterized both *in vitro* and *in vivo* for several FA subgroups. Direct evidence for mitochondrial dysfunction in FA cells was provided by Mukhopadhyay et al. (2006). In particular, it was shown that the FANCG protein is localized in the mitochondria and interacts with the mitochondrial peroxidase, peroxyredoxin 3 (PRDX3) and the G546R mutant of FANCG fails to interact physically with PRDX3 in contrasts to wild type. PRDX3 was deregulated in FA cells and they underwent cleavage by a calpain-like cysteine protease and mislocalization. These cells were also characterized by distorted mitochondrial structures and corresponding mitochondrial extracts had significantly decreased thioredoxin-dependent peroxidase activity. Overexpression of PRDX3 lowered the sensitivity of FAG cells to H_2O_2. In turn, downregulation of PRDX3 led to increased sensitivity to MMC. In fact, cells from FAA and FAC sub-types also revealed PRDX3 cleavage and decreased peroxidase activity (Mukhopadhyay et al. 2006).

Other reports documented mitochondrial dysfunction in FA cells (Bogliolo et al. 2002; Kumari et al. 2014; Rousset et al. 2002). They revealed an oxygen-dependent sensitivity of mitochondria and observed that in response to 8-methoxypsoralen photoreaction or ultraviolet A irradiation, FAA fibroblasts underwent mitochondrial matrix densification, unlike control fibroblasts. Bogliolo et al. (2002) demonstrated that mitochondrial function was challenged in FANCA- and FANCC-defective lymphoblastoid cells with rhodamine-1,2,3 and doxycycline. They observed that FA cells, unlike control cells, were not affected by these treatments, which resulted in lowered levels of ATP and in an increased number of cells undergoing apoptotic cell death. On the other hand, FA cells were very sensitive to 2-deoxy-d-glucose and iodoacetic acid, two inhibitors of the glycolytic metabolism and the findings suggested that FA cells had adapted to withstand mitochondrial stress while relying on glycolytic metabolism (Bogliolo et al. 2002). Several previous reports on the mitochondrial morphology in FA-A (Rousset et al. 1990), FA-D2 (Rousset et al. 2002), FA-C (Hadjur et al. 2001) and FA-G fibroblasts (Mukhopadhyay et al. 2006) suggested some possible changes in the structure of FA mitochondria. Figure 1 demonstrates an example of the common alterations in mitochondrial morphology seen in FAD2 cells, including wall ruptures, thinner walls and cristae, and smaller sizes (depicted by chevron signs). Mitochondria in FANCD2 complementation group suffer from similar abnormalities as in groups A and C and include low MMP and OCR, decreased ATP synthesis and increased ROS level (Kumari et al. 2014). Observation of distorted mitochondria in FA-G accompanied by mislocalization of PRDX3 has been also observed (Figure 2). Mitochondria from FA-G fibroblasts demonstrate frequent elongation and irregular shapes with acute and right angle branching (arrows). Some mitochondria have a z-shaped focal constriction of their diameters and bleb-like extensions. Interestingly, non-phosphorylable mutant cells (K561R FAD2) in which FANCD2 relocation to the DNA damaged sites (stall replication forks) have close to normal mitochondria (Figure

1) suggesting that FANCD2 pathway aiming to recruit post-chromatin FA-BRCA proteins (dysfunctional upon K561R mutation) does not directly significantly influence ROS-FA axis (Kumari et al. 2014).

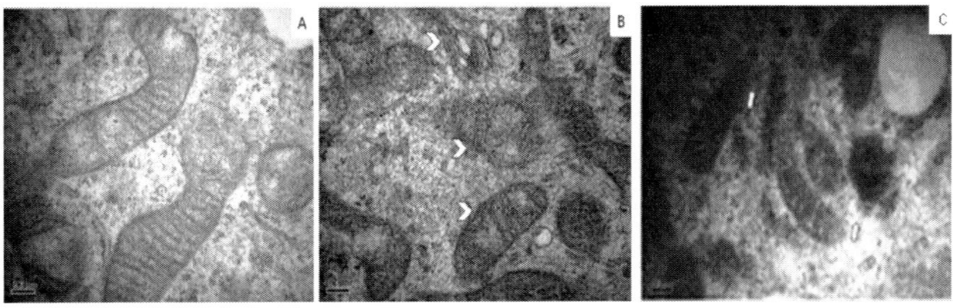

Figure 1. Mitochondrial abnormalities in FA cells.
Mitochondria of FANCD2 proficient (A), deficient (B) and corrected non-phosphorylable mutant K561 (C) cells. Irregular shaped mitochondria (shown by white arrows) with disrupted walls and cristae structures (chevron signs) are depicted. Resolution 28000.

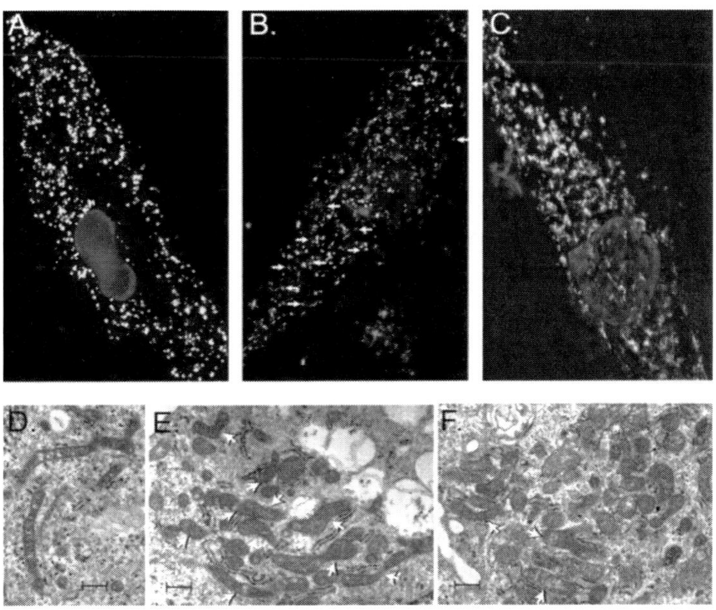

Figure 2. Mislocalization of PRDX3 and distorted mitochondrial structures in FA-G mutant fibroblasts. (A–C) Normal primary human fibroblasts (A), FA-G mutant (B), and FA-G–corrected (C) primary fibroblasts (PD352) were transfected with mitochondrial marker (pDsRed-Mito). The C-terminal PRDX3 antibody and FITC-labeled secondary antibody were used to detect the endogenous PRDX3 protein. Merge of the green and red images. Arrows (B) indicate the PRDX3 (green signals) that does not colocalize with the red mitochondrial marker in FA-G fibroblasts. (D and E) Transmission EM of FA-G mutant primary fibroblasts. Mitochondria from FA-G fibroblasts demonstrate frequent elongation and irregular shapes with acute and right angle branching (arrows). Some mitochondria have a z-shaped focal constriction of their diameters and bleblike extensions. (F) Transmission EM of FA-G–corrected primary fibroblasts with occasional mitochondria that have irregular outlines and branching pattern (top two arrowheads) and more normal structure (bottom arrowhead). Bars (D), 600 nm; (E and F) 400 nm. (Adapted from Mukhopadhyay et al. 2006).

Moreover, PRDX3 expression impacts sensitivity to OS and MMC independent of FANCD2 monoubiquitination (Figure 3). This supports the idea of existence of multiple dysfunctional pathways in FA-BRCA cascade, separated from classical DNA damage and repair. A study by Ravera et al. (2013) found a reduced oxygen consumption in FA-A cells when they were induced by pyruvate/malate, while respiration rate was similar in all samples when the cells were induced with succinate. This suggested a deficiency of mitochondrial complex I in FA-A cells. In addition to this, studies from our laboratories have demonstrated that FA cells have decreased oxygen uptake (Figure 4), high ROS and NOS levels (Figure 5) and low mitochondrial membrane potential (MMP) (Figure 6).

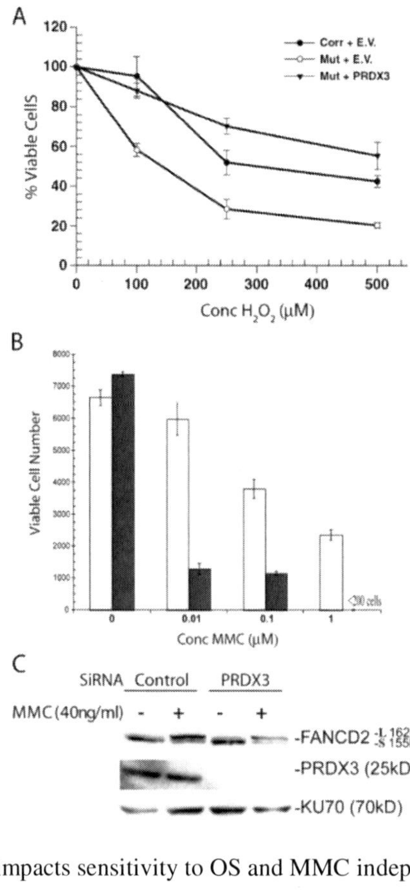

Figure 3. PRDX3 expression impacts sensitivity to OS and MMC independent of FANCD2 monoubiquitination.

(A) FA-G mutant (PD352-F1°) and corrected (PD352-F2) fibroblasts were transfected with the pEF6/His6 vector (empty vector [EV]) or pEF6-PRDX3 (PRDX3). 24 h later, cells were exposed to H_2O_2 for 90 min, and their viability was assessed with an MTT assay. A representative experiment is shown with the line going through the mean and error bars showing the minimum and maximum value for each cell line. (B) HeLa cells were transfected with control siRNA (white bars) or PRDX3 siRNA (black bars). Cells were exposed to MMC for additional 24 h, and the number of viable cells was measured by MTT assay. (C) HeLa cells were transfected with control or PRDX3 siRNA and then treated with 40 ng/ml MMC for 24 h. Protein lysates were immunoblotted with antibodies to PRDX3, FANCD2, and KU70. The long (L) and short (S) FANCD2 species are noted. (Adapted from Mukhopadhyay et al. 2006).

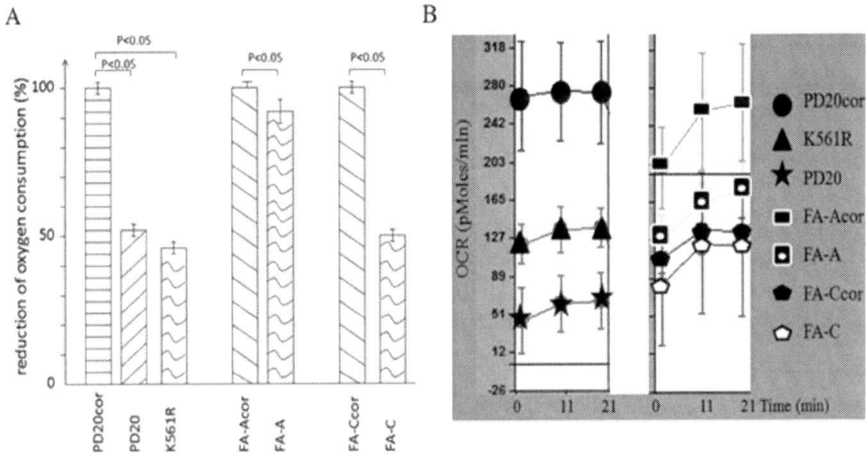

Figure 4. Reduction of oxygen uptake in FA cells.
(A) Bar diagram of oxygen uptake in FA deficient and proficient cells as measured in Figure 2D. Each bar is an average of three independent experiments. T-Student's test was applied. (B) Measurement of oxygen consumption by SeaHorse XF24 extracellular flux analyzer in FA proficient and deficient cells. Each time point represents results of 17 measurements (adapted from Kumari et al. 2014).

On a molecular level, this was in parallel with impaired response to ROS of major mitochondrial complexes and especially detoxifying enzyme superoxide dismutase 1 (SOD1) (Joenje et al. 1979; Yoshimitsu et al. 1984; Izakovic et al. 1985). As SOD1 removes superoxide radical (SOX), the lack of SOD1 may accumulate SOX in mitochondria. A measure of SOX levels revealed statistically significant SOX increase in siRNA FA-depleted cells. Overexpression of SOD1 in FA cells and a measure of critical mitochondrial functions by Seahorse XF24 revealed decreased Oxygen consumption rate (OCR) and coupled respiration capacity comparatively to corrected counterparts (Kumari et al. 2014).

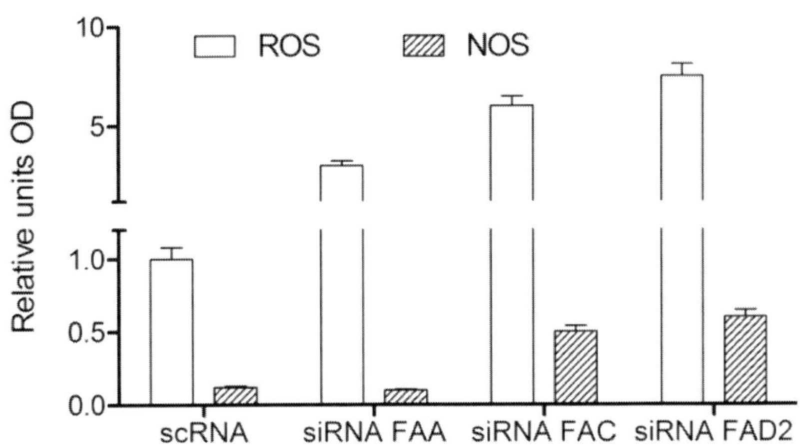

Figure 5. Elevated levels of ROS and NOS in FA-like cells.
HEK293 cells depleted for FANCA, FANCC, /FANCD2 by siRNA or treated with scrambled siRNA were processed for ROS measurement and the NOS measurement. Fluoremetric results of four replicas from two independent experiments are plotted as OD values.

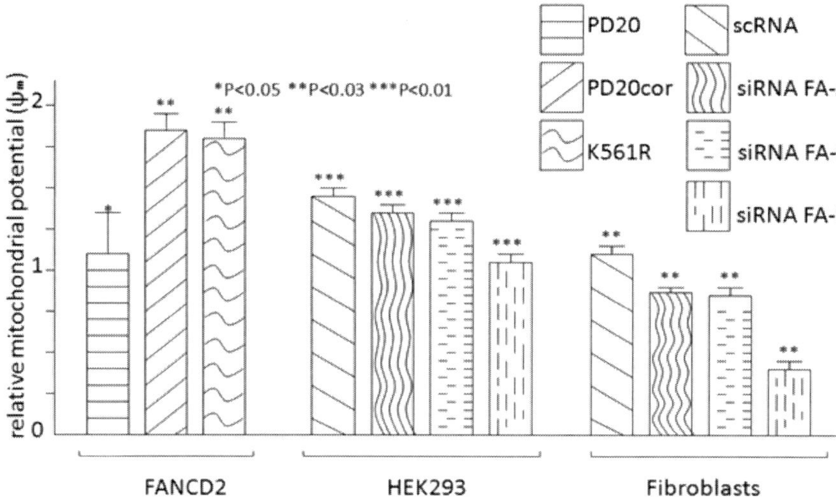

Figure 6. Measure of Mitochondrial membrane potential.
Relative Dcm values in FANCD2-deficient (PD20 or non-phosphorylable mutant K561R) or corrected (PD20cor) fibroblasts or FA siRNA transfected HEK293 or IMR90 lung fibroblasts growing in antibiotics-free Dulbecco's modified Eagle's media (DMEM) were measured spectrofotometrically from the JC-1 stained samples, and after normalization for the cell numbers were plotted onto diagram as fold differences. Experiments were independently performed four times and Student's t-test was applied to analyze the data. (adapted from Kumari et al. 2014).

In turn, SOD1 overexpression rescued OCR and coupled respiration capacity in FA cells and sensitivity of FA cells to MMC. We also demonstrated impaired response to H_2O_2 of both F1F0 ATPase (ATP synthase, Complex 5) and cytochrome C oxidase (COX, Complex 4) enzymes that are essential for the energy production. These results are indicative that some mitochondrial enzymes responsible for ROS protection were unable to respond to H_2O_2 in FA cells, suggesting impairment of mitochondrial detoxifying machinery in FA cells. Moderate decrease of Nitric Oxide Synthase (NOS) levels in FA-A, -C or –D2 siRNA depleted HEK293 was also detected. Treatment of the same cells with H_2O_2 further increased mitochondrial fragmentation. In turn, presence of ROS scavenger N-acetylcysteine (NAC) was able to lower overproduction of ROS and NOS in siRNA FA-depleted cells and decrease the number of cells with ROS-induced defragmented mitochondria (Kumari et al. 2014). Interestingly, FA-like cells depleted from mitochondria did not present high sensitivity to MMC suggesting that mitochondria-mediated ROS could be equally important for FA sensitivity to ROS (Figure 7) (Lyakhovich 2013). These findings demonstrate that FA cells have impaired MMP and disrupted mitochondrial chains, which is in parallel with the excess of intracellular ROS and abnormal mitochondrial morphology accompanied by the decreased ATP production and oxygen uptake (Kumari et al. 2014). This could be explained by the fact that excess of ROS may damage the FA mitochondria and result in an arrest of ATP synthesis, carbonylation and deactivation of various mitochondrial membrane proteins and also a decrease of MMP.

In order to disclose the relationship between the sensitivity of FA cells to inter cross-linking agents and the capacity of such agents to induce OS or alterations in the mitochondrial function of these cells, several experiments have been carried out. After the treatment of FA cell with MMC, the basal uncoupled (2 mg/mL oligomycin) or total respiration (2 mM FCCP) reveals the at least two-fold increase in respiratory state. However, the uncoupled respiration was higher in control cells (wild-type phenotype - HSC536 + FANCC) than in FA cells. The anti-oxidant agent Trolox, both in control and FA cells, neither did modify the uncoupled respiration rate nor changed in the activation of NF-κB (a marker of OS). Trolox delayed the loss of FA cell viability induced by MMC. The FA cellular death mediated by MMC involves the apoptotic pathways as evaluated by fluorescence microscopy of ethidium bromide and acridine orange. These results specify that OS and the apoptotic cell death induced by MMC in FA cells may be related to the distinct mitochondrial membrane permeability (unpublished data). Altogether, these observations indicate functional deficiencies of mitochondria in FA.

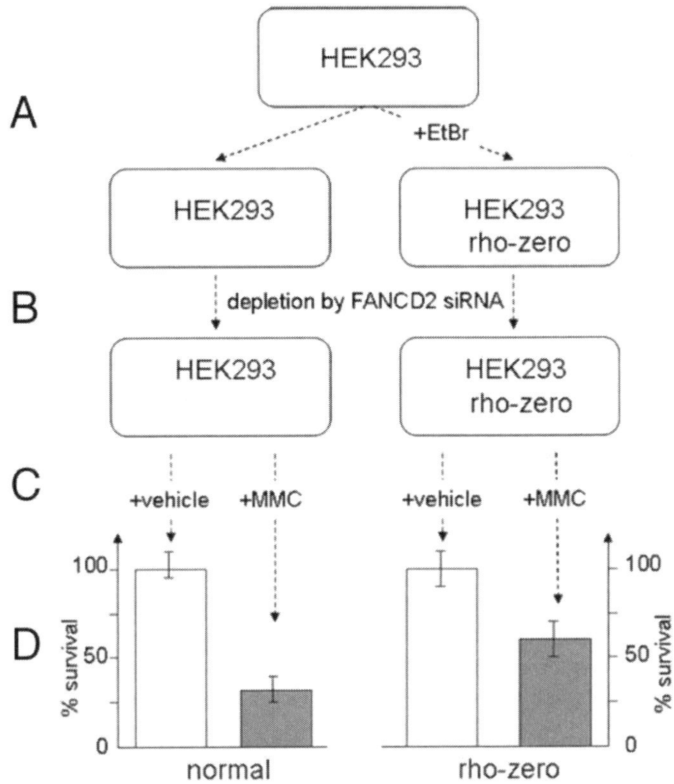

Figure 7. Role of Mitochondria in MMC sensitivity of FA cells.
FA-like cells with non-functional mitochondria are less sensitive to MMC. (A) Rho-zero cells were prepared by incubating HEK 293 cells with ethidium bromide. (B) Depletion of cells from FANC D2 has been performed by double transfection with FANC D2 siRNA. (C) Cells were incubated for 1 h with 0.5 uM MMC (final) or vehicle and MMC sensitivity assay was performed following 24 h post-treatment. (D) Cell sensitivity to MMC is increased for mitochondria-depleted rho-zero HEK293 cells vs. cells with functional mitochondria (adapted from Lyakhovich 2013).

Regulation of Mitochondrial ROS Production Antioxidants

Since mitochondria are a major source of intracellular ROS formation, they are in need of constant protection from the toxic action of these species. Such protection is provided by various low-molecular-weight antioxidants, as well as by multiple enzymatic defense systems (Sies 1997). Among the antioxidants, ubiquinone (Quiles et al. 2004) (Marriage et al. 2004) and vitamin E (Moore and Roberts 1998; Pfluger et al. 2004; Stephens et al. 1996) have been found to be particularly important. Selective delivery of synthetic antioxidants to mitochondria has been made possible by attachment of a lipophilic triphenylphosphonium cation to α-tocopherol (MitoVit E) or to ubiquinone (MitoQ) (Smith et al. 2003). The positive charge of these antioxidant derivatives enables them to accumulate in mitochondria because of the negative MMP inside these organelles and to provide significant protection from toxic OS. Further, recent studies suggest that dietary vitamin E supplementation can enrich hepatic mitochondria with protective levels of α-tocopherol in a dose dependent manner, and that this may prove beneficial in the treatment of OS–mediated liver disease, particularly when this is of mitochondrial origin (Sukalski et al. 1993; Fariss et al. 2001). Most important for mitochondrial antioxidant protection is the tripeptide glutathione, GSH (l-γ-glutamyl-l-cysteinylglycine), and multiple GSH-linked enzymatic defense systems (Orrenius et al. 2003; 2007). In fact, it was suggested that toxic cell death was often better correlated to depletion of the mitochondrial GSH pool than to overall intracellular GSH depletion.

Antioxidants and Their Effect on Mitochondrial Parameters in FA Cells

The pro-oxidant FA phenotype cells is characterized by alterations in the electron transport efficiency in the inner mitochondrial membrane with loss of efficiency of electron transfer from Complex I to Complex III which can in fact be restored by administration of exogenous Coenzyme Q (Ravera et al. 2013). As NAC decreases ROS production and normalizes mitochondrial respiration (Zafarullah et al. 2003) and NAC has already been reported to be only partially effective in limiting GSH depletion, particularly when employed in drugs targeted to mitochondria (Cuccarolo et al. 2012), it is expected that treatment of FA cells with NAC might increase oxygen consumption. Studies from Columbaro et al. (2014) have demonstrated that although treatment of FA cells with NAC was beneficial with restoration of biochemical parameters, the cellular structures did not improve and in fact it affected mitochondrial structure. The ability of NAC to restore Complex I activity is however inherently linked to an increase in ROS production and therefore in worsening of the already existent structural damage typical of FA-A cells. This is also suggestive of the existence of a primary damage to mitochondrial membranes in the FA-A phenotype.

There are some reports that lymphoblastoid cell lines derived from FA patients exhibit significant resistance to OS induced by treatments with H_2O_2 and various glutathione (GSH) inhibitors that induce production of ROS, GSH depletion and mitochondrial membrane depolarization. Among the various GSH inhibitors employed, FA cells appear particularly resistant to menadione (5µm) and ethacrynic acid (ETA, 50 µm), two drugs that specifically target mitochondria. Even after pre-treatment with buthionine sulfoximine, a GSH synthesis inhibitor that induces enhanced induction of ROS, FA cells maintain significant resistance to these drugs. These data suggest that the resistance to OS and the altered mitochondrial and metabolic functionality found in the FA mutant cells may indicate the survival strategy that is adopted in FA cells undergoing transformation. The study of redox and mitochondria regulation in FA may be of assistance in diagnosis of the disease and in the care of patients (Cuccarolo et al. 2012).

Regardless of being highly heterogeneous, the cells from all FA patients are hypersensitive to alkylating agents, in particular to diepoxybutane (DEB), and to oxidative DNA damage. Recent studies proved the defective mitochondria in FA cells, which is closely related to the overproduction of ROS and associated depletion of antioxidant defences, of which glutathione is a well-known biomarker. The evaluation studies revealed the presumed protective effect of a mitochondrial protective agent α-lipoic acid (ALA) and NAC, a direct antioxidant and a known precursor for glutathione synthesis, in spontaneous and DEB-induced chromosome instability (CI) in lymphocyte cultures from FA patients. The obtained results revealed that a cocktail of ALA and NAC can drastically improve the genetic stability in FA lymphocytes *in vitro*, depleted chromosomal instability (CI) by 60% and 80% in cultures from FA patients and FA mosaic/chimera patients, respectively. (Ponte et al. 2011; 2012). These results suggest that the combinatorial studies can be used as a prophylactic approach to delay progressive clinical symptoms in FA patients caused by CI, which can terminate in the delay of the progressive bone marrow failure and early cancer development.

Mitochondria and FA Pathology

It has been previously reported that cells with mutations in FA-A or FA-C rely, in part, on glycogenolytic metabolism, a strategy commonly associated with transformation and resistance of mutant FA-A and FA-C cells to cofactors interfering with the mitochondrial transport electron chain (Cappelli et al. 2013). Under physiologically normal conditions, in which mitochondria are active in ATP synthesis (State 3), the transport of ADP, phosphate and protons across the inner membrane accelerates the rate of electron transport; hence, most of the oxygen consumed by the respiratory chain is completely reduced to water. However, in pathological conditions, as in the case of FA, ROS accumulation affects mitochondrial functions, namely ATP production, ROS detoxification, and maintenance of MMP and OCR. It is likely that many FA mitochondria are shifted to a semi-resting state (State 4), where ATP production is defective and the rate of oxygen consumption is decreased. A high ROS level may lead to the failure of detoxifying enzymes to respond to ROS leading to mitochondrial abnormalities. Such vicious cycle corroborates well with the commonly accepted mitochondrial aging theory postulating that the ROS produced by the mitochondrial electron transport chain may further damage the mitochondria thus leading to more ROS production and amplifying mitochondrial damage. This theory does have some limitations. The most obvious one is that ROS amplification in FA defected mitochondria do not last indefinitely. This may suggest some neutralizing mechanism aiming either to cut the above circle or to stop production of ROS by specialized enzymes. Indeed, ROS overproduction and/or failure in ROS detoxification can result in pathogenic mechanisms. In turn, the presence of ROS scavengers may rescue mitochondrial function, thus increasing ATP production and decreasing ROS-dependent mitochondrial fragmentation. Oxidative stress from accumulated ROS has been associated with tumor formation (Schumacker 2006). Several studies suggest that ROS overproduction by extra mitochondrial source may be tumorigenic (Dröge 2002; Fridlyand and Philipson 2005); hence, therapies aimed at reducing ROS might offer effective means of combating malignancies (Kumar et al. 2008). The latter statement has some important clinical implications for FA cancer patients who require courses of chemotherapy or those undergoing conditioning before hematopoietic stem cell transplantation. Combinatorial treatment with chemotherapeutic drugs and proper ROS scavenging agents may decrease the toxic effects of DNA crosslinkers and hence benefit for better clinical outcome. At the same time, we should recall that simply adding antioxidative therapy is not enough. First, it is a palliative treatment as it will never correct the actual cause of FA. Secondly, ROS are important for proper metabolic functions of the cells. Significant decrease in ROS production will result in inability of a cell to struggle with infections (immune response based on lysosomal and peroxisomal degradation of targeted pathogens) or impairment of neurotransmitters' signaling processes. Therefore, antioxidant therapy should be applied with an extra care as it may create more adverse consequences than being a curative instrument. One of the possible solutions could be the application of some novel class of degradable antioxidants specifically targeting mitochondria. Regardless of the possible side-effects, mitochondria-mediated ROS in FA represents a challenging strategy. Eventually, mitochondrial activity and metabolism appear as a focal point in the pathological manifestations of FA. Knowledge of the pathogenic mechanisms involved in the resistance to apoptosis can help to develop new diagnostic and, hopefully, therapeutic strategies for this disease.

References

Alter, B. P. (1993). Fanconi's anaemia and its variability. *Br. J. Haematol.*, *85*, 9–14.

Auerbach, A. D. & Wolman, S. R. (1976). Susceptibility of Fanconi's anaemia fibroblasts to chromosome damage by carcinogens. *Nature, 261*, 494–496.

Bjelakovic, G., Nikolova, D. & Gluud, C. (2013). Antioxidant supplements to prevent mortality. *JAMA*, *310*, 1178–1179.

Bogliolo, M., Borghini, S., Abbondandolo, A. & Degan, P. (2002). Alternative metabolic pathways for energy supply and resistance to apoptosis in Fanconi anaemia. *Mutagenesis*, *17*, 25–30.

Brown, D. M., Donaldson, K., Borm, P. J., Schins, R. P., Dehnhardt, M., Gilmour, P., Jimenez, L. A. & Stone, V. (2004). Calcium and ROS-mediated activation of transcription factors and TNF-alpha cytokine gene expression in macrophages exposed to ultrafine particles. *Am. J. Physiol. Lung Cell Mol. Physiol.*, *286*, L344–53.

Cappelli, E., Ravera, S., Vaccaro, D., Cuccarolo, P., Bartolucci, M., Panfoli, I., Dufour, C. & Degan, P. (2013). Mitochondrial respiratory complex I defects in Fanconi anemia. *Trends Mol. Med.*, *19*, 513–514.

Clarke, A. A., Philpott, N. J., Gordon-Smith, E. C. & Rutherford, T. R. (1997). The sensitivity of Fanconi anaemia group C cells to apoptosis induced by mitomycin C is due to oxygen radical generation, not DNA crosslinking. *Br. J. Haematol., 96*, 240–247.

Columbaro, M., Ravera, S., Capanni, C., Panfoli, I., Cuccarolo, P., Stroppiana, G., Degan, P. & Cappelli, E. (2014). Treatment of FANCA cells with resveratrol and N-acetylcysteine: a comparative study. *PloS One*, *9*, e104857.

Cuccarolo, P., Viaggi, S. & Degan, P. (2012). New insights into redox response modulation in Fanconi's anemia cells by hydrogen peroxide and glutathione depletors. *FEBS J.*, *279*, 2479–2494.

Cumming, R. C., Lightfoot, J., Beard, K., Youssoufian, H., O'Brien, P. J. & Buchwald, M. (2001). Fanconi anemia group C protein prevents apoptosis in hematopoietic cells through redox regulation of GSTP1. *Nat. Med.*, *7*, 814–820.

Degan, P., Bonassi, S., De Caterina, M., Korkina, L. G., Pinto, L., Scopacasa, F., Zatterale, A., Calzone, R. & Pagano, G. (1995). In vivo accumulation of 8-hydroxy-2'-deoxyguanosine in DNA correlates with release of reactive oxygen species in Fanconi's anaemia families. *Carcinogenesis*, *16*, 735–741.

Dröge, W. (2002). Free radicals in the physiological control of cell function. *Physiol. Rev.*, *82*, 47–95.

Duchen, M. R. (2000). Mitochondria and calcium: from cell signalling to cell death. *J. Physiol. (Lond.), 529 Pt 1*, 57–68.

Dusre, L., Rajagopalan, S., Eliot, H. M., Covey, J. M. & Sinha, B. K. (1990). DNA interstrand cross-link and free radical formation in a human multidrug-resistant cell line from mitomycin C and its analogues. *Cancer Res.*, *50*, 648–652.

Fariss, M. W. & Zhang, J.-G. G. (2003). Vitamin E therapy in Parkinson's disease. *Toxicology*, *189*, 129–146.

Fariss, M. W., Nicholls-Grzemski, F. A., Tirmenstein, M. A. & Zhang, J. G. (2001). Enhanced antioxidant and cytoprotective abilities of vitamin E succinate is associated with a rapid uptake advantage in rat hepatocytes and mitochondria. *Free Radic. Biol. Med.*, *31*, 530–541.

Fariss, M. W., Chan, C. B., Patel, M., Van Houten, B. & Orrenius, S. (2005). Role of mitochondria in toxic oxidative stress. *Mol. Interv.*, *5*, 94–111.

De Flora, S., Izzotti, A., D'Agostini, F. & Balansky, R. M. (2001). Mechanisms of N-acetylcysteine in the prevention of DNA damage and cancer, with special reference to smoking-related end-points. *Carcinogenesis*, *22*, 999–1013.

Fridlyand, L. E. & Philipson, L. H. (2005). Oxidative reactive species in cell injury: Mechanisms in diabetes mellitus and therapeutic approaches. *Ann. N. Y. Acad. Sci.*, *1066*, 136–151.

Fridman, J. S. & Lowe, S. W. (2003). Control of apoptosis by p53. *Oncogene*, *22*, 9030–9040.

Futaki, M., Igarashi, T., Watanabe, S., Kajigaya, S., Tatsuguchi, A., Wang, J. & Liu, J. M. (2002). The FANCG Fanconi Anemia protein interacts with CYP2E1: possible role in protection against oxidative DNA damage. *Carcinogenesis*, *23*, 67–72.

Gupta, A., Rosenberger, S. F. & Bowden, G. T. (1999). Increased ROS levels contribute to elevated transcription factor and MAP kinase activities in malignantly progressed mouse keratinocyte cell lines. *Carcinogenesis*, *20*, 2063–2073.

Hadjur, S., Ung, K., Wadsworth, L., Dimmick, J., Rajcan-Separovic, E., Scott, R. W., Buchwald, M. & Jirik, F. R. (2001). Defective hematopoiesis and hepatic steatosis in mice with combined deficiencies of the genes encoding Fancc and Cu/Zn superoxide dismutase. *Blood*, *98*, 1003–1011.

Ishida, R. & Buchwald, M. (1982). Susceptibility of Fanconi's anemia lymphoblasts to DNA-cross-linking and alkylating agents. *Cancer Res.*, *42*, 4000–4006.

Izzotti, A., Bagnasco, M., Cartiglia, C., Longobardi, M., Camoirano, A., Tampa, E., Lubet, R. A., D & Flora, S. (2005). Modulation of multigene expression and proteome profiles by chemopreventive agents. *Mutat. Res., 591*, 212–223.

Joenje, H., Frants, R., Arwert, F., Bruin, G., Kostense, P., Kamp, J., Koning, J. & Eriksson, A. (1979). Erythrocyte superoxide dismutase deficiency in Fanconi's anaemia established by two independent methods of assay. *Scand. J. Clin. Lab. Invest., 39*, 759–764.

Joenje, H., Arwert, F., Eriksson, A. W., de Koning, H. & Oostra, A. B. (1981). Oxygen-dependence of chromosomal aberrations in Fanconi's Anaemia. *Nature*, *290*, 142-143.

Korkina, L. G., Samochatova, E. V., Maschan, A. A., Suslova, T. B., Cheremisina, Z. P. & Afanas'ev, I. B. (1992). Release of active oxygen radicals by leukocytes of Fanconi anemia patients. *J. Leukoc. Biol.*, *52*, 357–362.

Kruyt, F. A., Hoshino, T., Liu, J. M., Joseph, P., Jaiswal, A. K. & Youssoufian, H. (1998). Abnormal microsomal detoxification implicated in Fanconi anemia group C by interaction of the FAC protein with NADPH cytochrome P450 reductase. *Blood*, *92*, 3050–3056.

Kumar, B., Koul, S., Khandrika, L., Meacham, R. B. & Koul, H. K. (2008). Oxidative stress is inherent in prostate cancer cells and is required for aggressive phenotype. *Cancer Res.*, *68*, 1777–1785.

Kumari, U., Jun, W., Bay, B. & Lyakhovich, A. (2014). Evidence of mitochondrial dysfunction and impaired ROS detoxifying machinery in Fanconi anemia cells. *Oncogene*, *33*, 165–172.

Lee, J. H., Khor, T. O., Shu, L., Su, Z.-Y. Y., Fuentes, F. & Kong, A.-N. T. N. (2013). Dietary phytochemicals and cancer prevention: Nrf2 signaling, epigenetics, and cell death mechanisms in blocking cancer initiation and progression. *Pharmacol. Ther.*, *137*, 153–171.

Lyakhovich, A. (2013). Damaged mitochondria and overproduction of ROS in Fanconi anemia cells. *Rare Dis.*, *1*, e24048.

Lyakhovich, A. & Surrallés, J. (2010). Constitutive activation of caspase-3 and Poly ADP ribose polymerase cleavage in Fanconi anemia cells. *Mol. Cancer Res.*, *8*, 46–56.

Marriage, B. J., Clandinin, M. T., Macdonald, I. M. & Glerum, D. M. (2004). Cofactor treatment improves ATP synthetic capacity in patients with oxidative phosphorylation disorders. *Mol. Genet. Metab.*, *81*, 263–272.

Moore, K. & Roberts, L. J. (1998). Measurement of lipid peroxidation. *Free Radic. Res.*, *28*, 659–671.

Moustacchi, E., Diatlof &-Zito, C. (1985). DNA semi-conservative synthesis in normal and Fanconi anemia fibroblasts following treatment with 8-methoxypsoralen and near ultraviolet light or with X-rays. *Hum. Genet.*, *70*, 236–242.

Mukhopadhyay, S., Leung, K., Hicks, M., Hastings, P., Youssoufian, H. & Plon, S. (2006). Defective mitochondrial peroxiredoxin-3 results in sensitivity to oxidative stress in Fanconi anemia. *J. Cell Biol.*, *175,* 225-235.

Murphy, M. P. (2009). How mitochondria produce reactive oxygen species. *Biochem. J.*, *417*, 1–13.

Myung, S. K., Ju, W., Cho, B., Oh, S.-W. W., Park, S. M., Koo, B.-K. K. & Park, B. J. (2013). Efficacy of vitamin and antioxidant supplements in prevention of cardiovascular disease: systematic review and meta-analysis of randomised controlled trials. *Brit. Med. J.*, *346*, f10.

Nordenson, I. (1977). Effect of superoxide dismutase and catalase on spontaneously occurring chromosome breaks in patients with Fanconi's anemia. *Hereditas*, *86*, 147–150.

Orrenius, S., Zhivotovsky, B. & Nicotera, P. (2003). Regulation of cell death: the calcium-apoptosis link. *Nat. Rev. Mol. Cell Biol.*, *4*, 552–565.

Orrenius, S., Gogvadze, V. & Zhivotovsky, B. (2007). Mitochondrial Oxidative Stress: Implications for Cell Death. *Pharmacol. Toxicol.*, *47*, 143–183.

Pagano, G. & Youssoufian, H. (2003). Fanconi anaemia proteins: major roles in cell protection against oxidative damage. *Bioessays*, *25*, 589–595.

Pagano, G., Korkina, L. G., Degan, P., Del Principe, D., Lindau-Shepard, B., Zatterale, A. & Franceschi, C. (1997). In vitro hypersensitivity to oxygen of Fanconi anemia (FA) cells is linked to ex vivo evidence for oxidative stress in FA homozygotes and heterozygotes. *Blood*, *89*, 1111–1112.

Pagano, G., Degan, P., d' Ischia, M., Kelly, F. J., Nobili, B., Pallardó, F. V., Youssoufian, H. & Zatterale, A. (2005). Oxidative stress as a multiple effector in Fanconi anaemia clinical phenotype. *Eur. J. Haematol.*, *75*, 93–100.

Pagano, G., Talamanca, A. A., Castello, G., Pallardó, F. V., Zatterale, A. & Degan, P. (2012). Oxidative stress in Fanconi anaemia: from cells and molecules towards prospects in clinical management. *Biol. Chem.*, *393*, 11–21.

Pagano, G., Talamanca, A. A., Castello, G., Cordero, M. D., d'Ischia, M., Gadaleta, M. N., Pallardó, F. V., Petrović, S., Tiano, L. & Zatterale, A. (2014). Oxidative stress and mitochondrial dysfunction across broad-ranging pathologies: toward mitochondria-targeted clinical strategies. *Oxid. Med. Cell. Longev.*, *2014*, 541230.

Park, S.-J. J., Ciccone, S. L., Beck, B. D., Hwang, B., Freie, B., Clapp, D. W. & Lee, S.-H. H. (2004). Oxidative stress/damage induces multimerization and interaction of Fanconi anemia proteins. *J. Biol. Chem.*, *279*, 30053–30059.

Penketh, P. G., Hodnick, W. F., Belcourt, M. F., Shyam, K., Sherman, D. H. & Sartorelli, A. C. (2001). Inhibition of DNA cross-linking by mitomycin C by peroxidase-mediated oxidation of mitomycin C hydroquinone. *J. Biol. Chem.*, *276*, 34445–34452.

Pfluger, P., Kluth, D., Landes, N., Bumke-Vogt, C., Brigeliu &-Flohé, R. (2004). Vitamin E: underestimated as an antioxidant. *Redox Rep.*, *9*, 249–254.

Poljsak, B., Šuput, D. & Milisav, I. (2013). Achieving the balance between ROS and antioxidants: when to use the synthetic antioxidants. *Oxid. Med. Cell. Longev.*, *2013*, 956792.

Poll, E. H., Arwert, F., Joenje, H. & Eriksson, A. W. (1982). Cytogenetic toxicity of antitumor platinum compounds in Fanconi's anemia. *Hum. Genet.*, *61*, 228–230.

Ponte, F., Carvalho, F. & Porto, B. (2011). Protective effect of acetyl-l-carnitine and α-lipoic acid against the acute toxicity of diepoxybutane to human lymphocytes. *Toxicology*, *289*, 52–58.

Ponte, F., Sousa, R., Fernandes, A. P., Gonçalves, C., Barbot, J., Carvalho, F. & Porto, B. (2012). Improvement of genetic stability in lymphocytes from Fanconi anemia patients through the combined effect of α-lipoic acid and N-acetylcysteine. *Orphanet J. Rare Dis.*, *7*, 28.

Pritsos, C. A. & Sartorelli, A. C. (1986). Generation of reactive oxygen radicals through bioactivation of mitomycin antibiotics. *Cancer Res.*, *46*, 3528–3532.

Quiles, J. L., Ochoa, J. J., Huertas, J. R. & Mataix, J. (2004). Coenzyme Q supplementation protects from age-related DNA double-strand breaks and increases lifespan in rats fed on a PUFA-rich diet. *Exp. Gerontol.*, *39*, 189–194.

Rani, R., Li, J. & Pang, Q. (2008). Differential p53 engagement in response to oxidative and oncogenic stresses in Fanconi anemia mice. *Cancer Res.*, *68*, 9693–9702.

Ravera, S., Vaccaro, D., Cuccarolo, P., Columbaro, M., Capanni, C., Bartolucci, M., Panfoli, I., Morelli, A., Dufour, C., Cappelli & E.,Degan P. (2013). Mitochondrial respiratory chain Complex I defects in Fanconi anemia complementation group A. *Biochimie*, *95*, 1828–1837.

Reuter, T. Y., Medhurst, A. L., Waisfisz, Q., Zhi, Y., Herterich, S., Hoehn, H., Gross, H. J., Joenje, H., Hoatlin, M. E., Mathew, C&G., et al. (2003). Yeast two-hybrid screens imply involvement of Fanconi anemia proteins in transcription regulation, cell signaling, oxidative metabolism, and cellular transport. *Exp. Cell Res.*, *289*, 211–221.

Roomi, M. W., Kalinovsky, T., Roomi, N. W., Niedzwiecki, A. & Rath, M. (2012). In vitro and in vivo inhibition of human Fanconi anemia-associated head and neck squamous cell carcinoma by a novel nutrient mixture. *Int. J. Oncol.*, *41*, 1996–2004.

Rousset, S., Nocentini, S., Revet, B. & Moustacchi, E. (1990). Molecular analysis by electron microscopy of the removal of psoralen-photoinduced DNA cross-links in normal and Fanconi's anemia fibroblasts. *Cancer Res.*, *50*, 2443–2448.

Rousset, S., Nocentini, S., Rouillard, D., Baroche, C. & Moustacchi, E. (2002). Mitochondrial alterations in fanconi anemia fibroblasts following ultraviolet A or psoralen photoactivation. *Photochem. Photobiol.*, *75*, 159–166.

Ruppitsch, W., Meisslitzer, C., Weirich-Schwaiger, H., Klocker, H., Scheidereit, C., Schweiger, M., Hirsc &-Kauffmann, M. (1997). The role of oxygen metabolism for the pathological phenotype of Fanconi anemia. *Hum. Genet.*, *99*, 710–719.

Saadatzadeh, M. R., Bijangi-Vishehsaraei, K., Hong, P., Bergmann, H. & Haneline, L. S. (2004). Oxidant hypersensitivity of Fanconi anemia type C-deficient cells is dependent on a redox-regulated apoptotic pathway. *J. Biol. Chem.*, *279*, 16805–16812.

Schindler, D. & Hoehn, H. (1988). Fanconi anemia mutation causes cellular susceptibility to ambient oxygen. *Am. J. Hum. Genet.*, *43*, 429–435.

Schumacker, P. T. (2006). Reactive oxygen species in cancer cells: live by the sword, die by the sword. *Cancer Cell*, *10*, 175–176.

Shyamsunder, P., Vidyasekar, P., Shukla, A. R., Mohan, S. & Verma, R. S. (2013a). Lowered expression levels of a tumor suppressor gene - caveolin-1 within dysregulated gene networks of Fanconi anemia. *Gene*, *527*, 521–528.

Shyamsunder, P., Ganesh, K. S., Vidyasekar, P., Mohan, S. & Verma, R. S. (2013b). Identification of novel target genes involved in Indian Fanconi anemia patients using microarray. *Gene*, *531*, 444–450.

Sies, H. (1997). Oxidative stress: oxidants and antioxidants. *Exp. Physiol.*, *82*, 291–295.

Smith, R. A., Kelso, G. F., Blaikie, F. H., Porteous, C. M., Ledgerwood, E. C., Hughes, G., James, A. M., Ross, M. F., Asin-Cayuela, J., Cochemé, H. M., Filipovska, A. & Murphy,

M. P. (2003). Using mitochondria-targeted molecules to study mitochondrial radical production and its consequences. *Biochem. Soc. Trans.*, *31*, 1295–1299.

Stephens, N. G., Parsons, A., Schofield, P. M., Kelly, F., Cheeseman, K. & Mitchinson, M. J. (1996). Randomised controlled trial of vitamin E in patients with coronary disease: Cambridge Heart Antioxidant Study (CHAOS). *Lancet*, *347*, 781–786.

Sukalski, K. A., Pinto, K. A. & Berntson, J. L. (1993). Decreased susceptibility of liver mitochondria from diabetic rats to oxidative damage and associated increase in alpha-tocopherol. *Free Radic. Biol. Med.*, *14*, 57–65.

Szybalski, W. & Iyer, V. N. (1964). Crosslinking of dna by enzymatically or chemically activated mitomycins and porfiromycins, bifunctionally "alkylating" antibiotics. *Fed. Proc.*, *23*, 946–957.

Taniguchi, T. & D'Andrea, A. D. (2006). Molecular pathogenesis of Fanconi anemia: recent progress. *Blood*, *107*, 4223–4233.

Turrens, J. F. (2003). Mitochondrial formation of reactive oxygen species. *J. Physiol. (Lond.)*, *552*, 335–344.

Wang, X. & D'Andrea, A. D. (2004). The interplay of Fanconi anemia proteins in the DNA damage response. *DNA Repair (Amst.)*, *3*, 1063–1069.

Zafarullah, M., Li, W. Q., Sylvester, J. & Ahmad, M. (2003). Molecular mechanisms of N-acetylcysteine actions. *Cell. Mol. Life Sci.*, *60*, 6–20.

Zanier, R., Briot, D., Dugas du Villard, J.-A. A., Sarasin, A. & Rosselli, F. (2004). Fanconi anemia C gene product regulates expression of genes involved in differentiation and inflammation. *Oncogene*, *23*, 5004–5013.

Zhang, J. G., Nicholls-Grzemski, F. A., Tirmenstein, M. A. & Fariss, M. W. (2001). Vitamin E succinate protects hepatocytes against the toxic effect of reactive oxygen species generated at mitochondrial complexes I and III by alkylating agents. *Chem. Biol. Interact.*, *138*, 267–284.

Zhang, Q. S., Eaton, L., Snyder, E. R., Houghtaling, S., Mitchell, J. B., Finegold, M., Van Waes, C. & Grompe, M. (2008). Tempol protects against oxidative damage and delays epithelial tumor onset in Fanconi anemia mice. *Cancer Res.*, *68*, 1601–1608.

Zhang, Q. S., Marquez-Loza, L., Sheehan, A. M., Watanabe-Smith, K., Eaton, L., Benedetti, E., Major, A., Schubert, K., Deater, M., Joseph, E. & Grompe, M. (2014). Evaluation of resveratrol and N-acetylcysteine for cancer chemoprevention in a Fanconi anemia murine model. *Pediatr. Blood Cancer*, *61*, 740–742.

In: Fanconi Anemia and Oxidative Stress
Editor: Giovanni Pagano
ISBN: 978-1-63482-297-8

Chapter 5

Aldehydes As Markers of Redox Imbalance in Fanconi Anemia and in Other Oxidative Stress-Related Disorders

Paola Manini*[*] *and Marco d'Ischia
Department of Chemical Sciences, University of Naples Federico II, Naples, Italy

Abstract

Redox imbalance in biological systems is associated with a complex interplay of oxidative pathways leading to more or less extensive degradation and breakdown of important cellular constituents including chiefly lipids, such as polyunsaturated fatty acids (PUFAs), as well as carbohydrates and proteins, among others. PUFA oxidation is a central event in lipid peroxidation and is associated with a chain reaction involving peroxide intermediates. These latter eventually break down to give α,β-unsaturated aldehydes, including 4-hydroxy-2-nonenal (HNE), acrolein, crotonaldehyde, and dialdehydes, such as malondialdehyde, glyoxal and methylglyoxal. Oxidative stress-related aldehydes (OSRAs) are currently recognized to play a complex and crucial role in the onset and progression of several disease states, including vascular diseases, neurodegeneration and cancer. An important role of aldehydes in lipofuscinogenesis and age-related disorders has also been demonstrated. Besides being the measurable end product of lipid peroxidation and oxidative cell injury, aldehydes can act themselves as amplifiers of damaging events through covalent reaction with cellular targets such as proteins and nucleic acids. HNE, for example, has been implicated in a variety of cytopathological effects associated to oxidative stress and is currently one of the best recognized and most studied cytotoxic products of lipid peroxidation. The ability of glyoxal to cause interstrand DNA crosslinks is another remarkable example for the active role of aldehydes in mediating toxic mechanism in disease states, and the activation of the glyoxalase system to control glyoxal-mediated cell damage is indirect evidence of the

[*]Corresponding Author: E-mail: pmanini@unina.it.

impact of these processes on the progression of disease. Several conditions have been associated with an increase in the levels of endogenous aldehydes, including Fanconi anaemia, and the impact of this phenomenon may vary from one disease to the other depending on levels and potential targets involved, as well as the underlying specific pathways of toxicity. Starting from an initial survey of the basic mechanisms of aldehyde generation from lipid peroxidation and other oxidative processes following the onset of redox imbalance, this chapter aims at addressing a number of central issues in oxidative stress-related disorders: 1) what are the mechanisms of OSRAs toxicity in the progression and final outcome of the disease? 2) What are the main targets of OSRAs and what is the role of these interactions in the etiopathogenesis of oxidative stress–related disease states? 3) Can OSRA quantitation yield clues about the advancement and outcome of diseases? Far from being comprehensive, the present survey seeks rather to provide an inspiring and critical perspective of the biomedical relevance of aldehydes as markers of redox imbalance and their potential diagnostic and prognostic value with special reference to Fanconi anaemia and rare cancer-prone diseases.

Keywords: α,β-unsaturated aldehydes, oxidative stress, lipid peroxidation

Introduction

Reactive oxygen species (ROS) are continuously generated within living systems and the inability to manage ROS load leads to elevated oxidative stress and cell damage. Oxidative stress is coupled to the oxidative degradation of lipid membranes, enzymes, proteins and polysaccharides. These processes generate over 400 types of aldehydes, many of which are highly reactive and toxic for their ability to form Schiff base adducts with amine groups such as terminal or epsilon groups of lysine and arginine, and thioester bonds with nucleophilic cysteine residues. The reactivity of aldehydes is, however, further enhanced by α,β-unsaturation. Such aldehydes are abundant products of combustion and lipid peroxidation. The presence of a conjugated system of double bonds makes aldehydes more prone to react with non-protein and protein thiols and amines; furthermore, α,β-unsaturated aldehydes can readily cross-link proteins, lipids, and nucleic acids through Michael addition and Schiff base formations. This dual reactivity is a central feature of the biological activity of such aldehydes and is thought to represent the underlying mechanism of their toxicity (Jacobs and Marnett 2010; Voulgaridoua et al. 2011).

The aldehyde motif is present in a very large number of biological compounds. These include glucose and its metabolic intermediates (e.g., glyoxal and methylglyoxal), lipid peroxidation products (e.g., 4-hydroxy-*trans*-2-nonenal, HNE; 1-palmitoyl-2-(5-oxovaleroyl)-*sn*-glycero-3-phosphatidyl choline, POVPC), metabolites of neurotransmitters (e.g., succinic semi-aldehyde, 3,4-dihydroxyphenylglycoaldehyde, DOPGAL, 3-methoxy,4-hydroxyphenyl-glycoaldehyde, MOPGAL), products of DNA oxidation (e.g., base propenals), and enzymatic end products (e.g., acrolein derived from polyamine oxidase or semicarbazide-sensitive amine oxidase activity). In addition, living systems are exposed to aldehydes from several exogenous sources, including food and beverages.

Herein we report on the oxidative-stress related pathways of α,β-unsaturated aldehyde formation and their mechanisms of toxicity; moreover, an overall view of the role of α,β-unsaturated aldehydes in the onset of oxidative stress-induced diseases will be provided too.

Oxidative-Stress Related Pathways of α,β-Unsaturated Aldehyde Formation

Certainly lipid peroxidation represents the main oxidative stress related pathway leading to an over-production of α,β-unsaturated aldehydes (Zimniak 2011; Marnett et al. 2003); but there are also other pathways contributing to the formation of these highly reactive species, such as the sugar and polyol degradation and the amino acids/proteins and nucleic acids oxidation pathways (Wolff et al. 1991; Wautier and Schmidt 2004). Under physiological conditions these events are under enzymatic control preventing the accumulation of toxic species; but under oxidative stress conditions, the cellular defence system is impaired so that free radical chain reactions grow, ROS and RNS accumulate and the formation of reactive byproducts, such as α,β-unsaturated aldehydes, is stressed. Among OSRAs, the most important for occurrence and relevance in cito- and genotoxic processes are: acrolein (ACR), crotonaldehyde (CRA), malondialdehyde (MDA), 4-hydroxy-*trans*-2-nonenal (HNE), glyoxal (GLX) and methylglyoxal (MGLX) (2).

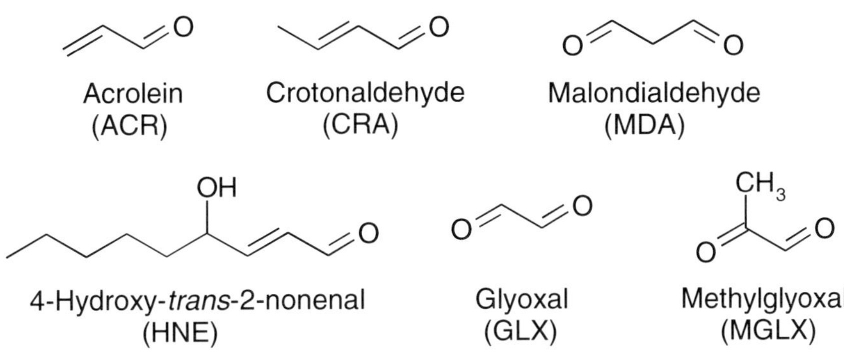

Acrolein (ACR)

ACR can be produced from peroxidation of membrane phospholipids, although the major aldehydes produced during lipid peroxidation are malondialdehyde and HNE (Uchida 1999), but it is also formed endogenously through metabolism of amino acids and polyamines (Table 1). As a matter of fact, ACR has been recognized as the major toxic compound produced from spermine and spermidine by amine oxidase in uremic subjects (Sakata et al. 2003; Igarashi et al. 2006). *In vivo*, ACR is also known to be formed from threonine by neutrophil myeloperoxidase at sites of inflammation (Anderson et al. 1997), and from methionine (Stevens and Maier 2008). Drug metabolism is another source of ACR. It is well known, in fact, that the oxazaphosphorines (cyclophosphamide, ifosfamide, and trofosfamide), although widely used in clinical practice for their antitumor and immunomodulatory activities, are responsible for urotoxicity, neurotoxicity, and nephrotoxicity. This is because their metabolism, involving cytochrome P450 biotransformations, leads to the formation of ACR (Hebert et al. 2010) which has been associated with the development of hemorrhagic cystitis or diffuse inflammation of the bladder resulting in dysuria, hematuria, and hemorrhage.

ACR and many other type-2 alkenes constitute a large family of environmental and food contaminants that include acrylamide, methyl vinyl ketone, methyl acrylate, and HNE, all characterized by electrophilic properties (LoPachin et al. 2008). ACR is indeed a common air pollutant, to which humans are continuously exposed and is present in high concentrations in wood, cotton, and tobacco smoke (10–140 µg/cigarette), automobile exhaust and industrial waste and emissions. ACR, as one of the major toxic by-products of smoke developed during pyrolysis or combustion of plastic materials, has been linked to severe pulmonary damage and recognized as one of the main causes of death following smoke inhalation (Einhorn 1975).

Relatively high levels of ACR have been detected in beer, wine, rum, bread, and other foods (Feron et al. 1991). The formation of ACR in foods, especially cooking oils, is further increased by cooking, frying, and reheating (Fullana et al. 2004).

ACR has a strong electrophilic reactivity towards nucleophiles; therefore, it disrupts the redox control of protein function and causes cytotoxicity via irreversible adduction. In addition, ACR may play a role in the pathogenesis of cardiovascular and neurodegenerative disorders (Korkmaz et al. 2007). It is an important oxidative stress biomarker for lipid peroxidation and ACR levels increase during aging and in disease, such as atherosclerosis and Alzheimer's disease (Batista et al. 2007; Conklin et al. 2009). Several recent studies have linked ACR exposure to atherosclerosis (Conklin et al. 2009), hypertension (LoPachin et al. 2008), dyslipidemia (Faroon et al. 2008), and infarction (Einhorn 1975).

Crotonaldehyde (CRA)

CRA is abundant in the environment and is also produced endogenously during lipid metabolism (Table 1) (Stone et al. 2008). CRA is reported to be present in many foods, such as fish, meat, fruit, and vegetables, and in various liquors (Eder and Budiawan 2001). It is formed as a product of lipid peroxidation and is also produced during the combustion of carbon-containing fuels and other materials (Kawaguchi-Niida et al. 2006); cigarette smoke is another important source of CRA (31–169 µg/kg body weight) (Eder et al. 1999). CRA is mutagenic without metabolic activation in numerous cell systems (Neudecker et al. 1989) and induces hepatic tumors in rodents (Eder et al. 1999). The toxicity of CRA is caused by its strongly reactive electrophilic carbonyl group (Esterbauer et al. 1991). Many studies have indicated that CRA directly or enzymatically conjugates with glutathione (GSH) thereby reducing the GSH levels (Van Der Toorn et al. 2007). Previous studies showed that CRA can modulate biological reactions through various downstream signaling pathways and cause cellular oxidative stress (Liu et al. 2010).

Malondialdehyde (MDA)

Several hypotheses describing the formation of MDA *in vivo* have been proposed. As proposed by Pryor and Stanley (1975) and later confirmed by Frankel and Neff (1983), oxidized lipids are able to produce MDA as a decomposition product (Table 1). The mechanism is thought to involve formation of prostaglandin-like endoperoxides from polyunsaturated fatty acids with two or more double bonds. An alternative mechanism presented by Esterbauer et al. (1991) is based on successive hydroperoxide formation and β-cleavage of polyunsaturated fatty acids. MDA is then directly formed by β-scission of a 3-hydroperoxyaldehyde or by reaction between acrolein and hydroxyl radicals. While oxidation of polyunsaturated fatty acids is the major source of MDA *in vivo*, other minor sources exists such as byproducts of free radical generation by ionizing radiation and of the biosynthesis of prostaglandins (Hecker and Ullrich 1989).

4-Hydroxy-Trans-2-Nonenal (HNE)

HNE is a lipid peroxidation end-product of enzymatic transformation of *n*-6 PUFAs (arachidonic acid, linoleic acid and other) by 15-lipoxygenases (15-LOX) (Table 1). The main precursors of HNE in human are 13-hydroperoxyoctadecadienoic acid (13-HPODE) produced by the oxidation of linoleic acid by 15-LOX-1 (Schneider et al. 2001) and 15-hydroperoxyeicosatetraenoic acids (15-HPETE) produced by the oxidation of arachidonic acid by 15-LOX-2 (Riahi et al. 2010).

Spickett (2013) recently reviewed the mechanisms of formation of HNE during lipid peroxidation and showed that the main processes leading to HNE are likely β-cleavage reaction of lipid alkoxy-radicals, which can be summarized into five generic mechanisms: (i) reduction of the hydroperoxide to a lipid alkoxy radical by transitionmetal ions, such as Fe^{2+} followed by β-scission; (ii) protonation of the lipid hydroperoxide yields an acidified lipid hydroperoxide that undergoes Hock rearrangement of a C–C to C–O bond followed by hydrolysis and Hock cleavage; (iii) the lipid peroxyl radical of the hydroperoxides permits their facile cyclization to dioxetane and ending with dioxetane cleavage; (iv) free radical attack to ω-6 PUFA on bis-allyl site yielding a free radical intermediate, that further reacts with molecular oxygen to generate hydroperoxide derivatives such as 13-HPODE or 15-HPETE. The abstraction of an allylic hydrogen of their structure produce another radical intermediate that after oxygenation step forms the corresponding dihydroperoxyde derivative (unstable), which after Hock rearrangement and cleavage produces 4-hydroperoxy-2E-nonenal (4SHPNE), immediate precursor of HNE; and (v) the oxidation products generated after reaction of linoleate-derived hydroperoxy epoxide (13-Hp-Epo-Acid) with Fe^{+2} yields an alkolxyl radical, which after β-scission yields different aldehydes including HNE.

HNE, one of the more reactive α,β-unsaturated aldehyde, is a known marker of oxidative stress. HNE is a potent alkylating agent that reacts with DNA and proteins, thereby generating various types of adducts (Esterbauer et al. 1991; Uchida et al. 1999). These adducts can induce stress signaling pathways and apoptosis (Kode et al. 2006). It has been reported that cigarette smoke extract (CSE) causes HNE production either directly or indirectly via lipid peroxidation in various cell types.

In another study by Kode (2006), CSE caused a dose-dependent increase in oxidative stress in various cell lines and in HNE levels in small airway epithelial cells (SAECs). CSE-induced cytotoxicity in different cell lines has been attributed to an increase in the endogenous production of HNE.

Kumagai showed that HNE may be a major inflammatory mediator in the development and progression of atherogenesis (Kumagai et al. 2004). HNE is reported to be producing nerve terminal toxicity by forming adducts that play a critical role in Alzheimer's disease (Butterfield et al. 2010). In addition, studies have revealed that HNE is associated with several other pathological conditions, such as acute respiratory distress syndrome (Rahman et al. 2002) and atherosclerosis (Napoli et al. 1997).

Glyoxal (GLX) and Methylglyoxal (MGLX)

Glyoxal (GLX) and methylglyoxal (MGLX) originate from both exogenous and endogenous pathways (Table 1). Exogenous sources are primarily dietary, and these include beer, wine, tea, coffee, yogurt, bread, rice, soybean paste, soy sauce, and oil (Kielhorn et al. 2004). Foods that have been fermented, roasted, baked, or fried are also a rich source of GLX and MGLX. Environmental sources of GLX and MGLX include cigarette smoke, smoke from residential log fires, vehicle exhaust, smog, fog, and some household cleaning products. It may also be introduced into the air during combustion of various materials. It has been found in samples of soil, groundwater, seawater, and sediment.

Whether GLX and MGLX from these sources is absorbed into the body through the gut, lungs, or skin to significantly affect GLX and MGLX levels in tissues is not known with certainty. In addition to the many exogenous sources, GLX and MGLX are endogenously produced via autoxidation of carbohydrates and ascorbate and via the degradation of glycated proteins. GLX is also produced during lipid peroxidation and the oxidative metabolism of nucleic acids, whereas MGLX is also formed during the oxidative metabolism of threonine and acetone (Lange et al. 2012; Abordo et al. 1999).

Reactions of Oxidative Stress-Related Aldehydes with Biological Targets and Other Mechanisms of Cytotoxicity

The biochemical routes of OSRAs metabolism can be summarized according to stress levels. Under physiological or low stress levels the major detoxification steps are: a) enzymatic conjugation with GSH; b) reduction by the action of enzymes of the family of the aldose reductases (AR); c) oxidation by the action of enzymes of the family of the aldehyde dehydrogenase (ALDH). At high stress levels, OSRAs accumulate and form covalent adducts with protein, enzymes and nucleic acids (Figure 1) (Spickett 2013).

As a bifunctional electrophile, OSRAs are highly reactive species. OSRAs reactivity is mainly based on their electrophilicity making them strongly reactive toward nucleophiles, such as basic amino acid residues (i.e., lysine, histidine, or arginine). Initial reactions between OSRAs and free amino acids or protein generate Schiff-base adducts (Esterbauer et al. 1991;

Pizzimenti et al. 2013; Traverso et al. 2004). OSRAs can also participate in secondary deleterious reactions by promoting intramolecular or intermolecular protein/DNA crosslinking that may induce profound alteration in the biochemical properties of biomolecules and accumulate during aging and in chronic diseases (see Figure 2 for some representative exemples of OSRAs-deoxyguanosine adducts) (Negre-Salvayre et al. 2008; Ayala et al. 2014; Slatter et al. 2004; Cheng et al. 2011).

Table 1. Main endogenous and exogenous sources of OSRAs

OSRA	Source	Ref.	OSRA	Source	Ref.
ACR	Lipid peroxidation	[7]	GLX	Monosaccharides	[48,49]
	Amino acids	[10,11]		Advanced glycation end-products	[50]
	Metabolism of polyamines	[8,9]		Lipid peroxidation	[51-53]
	Combustion (cigarette smoke, smog)	[14]		Nucleic acid oxidation	[54,55]
	Foods (mainly after frying and reheating) and beverages (beer, wine, rum)	[15,16]		Foods (bread, rice, soybean sauce) and beverages (wine tea and coffee)	[44]
CRA	Lipid peroxidation	[22]	MGLX	Monosaccharides	[48,49, 56]
	Combustion (cigarette smoke and smog)	[24,25]		Advanced glycation end-products	[50]
	Foods (fish, meat, fruit) and beverages (liquors)	[23]		Metabolism of acetone	[57]
MDA	Lipid peroxidation	[47]		Metabolism of threonine	[58]
HNE	Lipid peroxidation	[34,35]		Foods (bread, rice, soybean sauce) and beverages (wine tea and coffee)	[44]

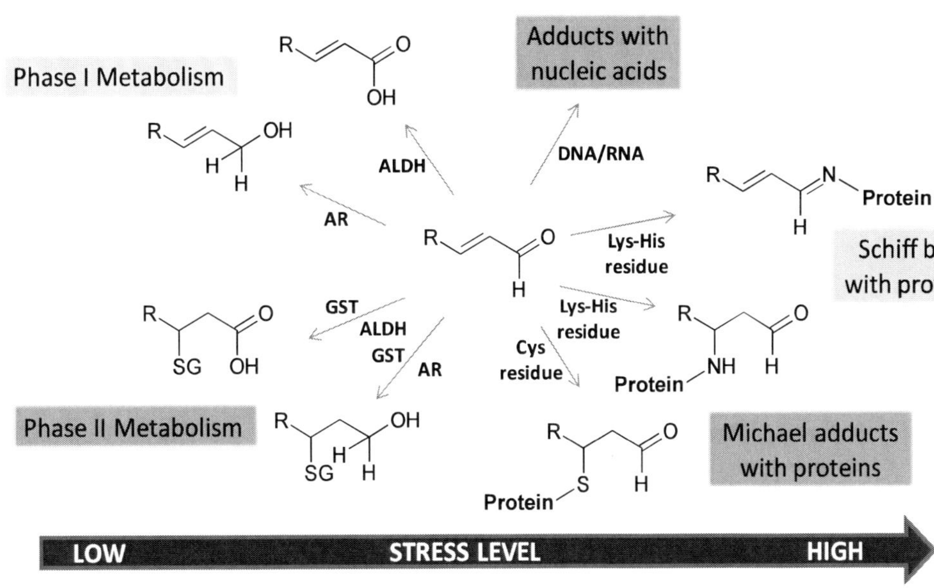

Figure 1. Main metabolic pathways of OSRAs. ALDH, aldehyde dehydrogenase; AD, aldehyde reductase; GST, glutathionyl-*S*-transferase.

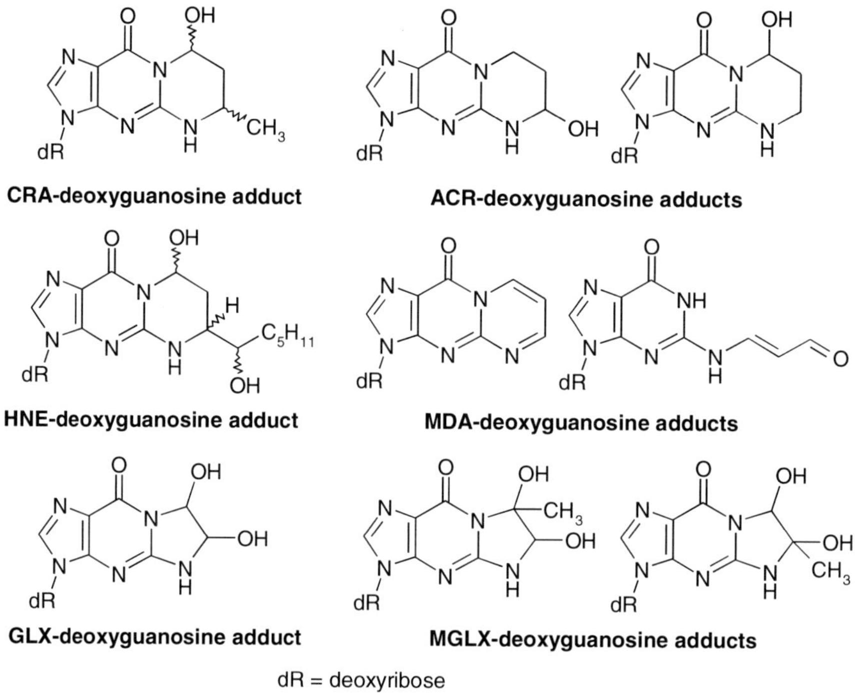

Figure 2. Main OSRAs deoxyguanosine adducts.

A recent review shows a list of up to thirty-three proteins known to be modified by MDA and including enzymatic proteins, carrier proteins, cytoskeletal proteins, and mitochondrial and antioxidant proteins (Zarkovic et al. 2013). It has also been proposed that MDA could react with several nucleosides to form adducts to deoxyguanosine and deoxyadenosine (Niedernhofer et al. 2003; Del Rio et al. 2005; Marnett 1999; 2002), becoming an important contributor to DNA damage and mutation (Niedernhofer et al. 2003; VanderVeen et al. 2003). In the absence of repair, MDA-DNA adducts may lead to mutations (point and frameshift) (VanderVeen et al. 2003), strand breaks (Niedernhofer et al. 2003; Voehringer et al. 1998), cell cycle arrest (Ji et al. 1998), and induction of apoptosis (Willis et al. 2004). This MDA-induced DNA alteration may contribute significantly to cancer and other genetic diseases.

HNE, which has been subjected to intensive scientific scrutiny in 1990s, is considered as one of the major toxic products generated from lipid peroxides (Esterbauer et al. 1991). HNE acts both as signaling molecule and as cytotoxic product of lipid peroxidation causing long-lasting biological consequences, in particular by covalent modification of macromolecules. HNE is an extraordinarily reactive compound containing three functional groups: (i) C=C double bond that can be target for Michael additions, mainly driven by thiols, reduction or epoxidation, (ii) carbonyl group which can yield acetal/thioacetal or can be target for Schiff-base formation, oxidation, or reduction, and (iii) hydroxyl group which can be oxidized to give a ketone (Schaur 2003).

HNE is the most intensively studied lipid peroxidation end-product (Dalleau et al. 2013; Balogh and Atkins 2011; Siems and Grune 2003; Alary et al. 2003), in relation not only to its physiological and protective function as signaling molecule stimulating gene expression, but also to its cytotoxic role inhibiting gene expression and promoting the development and progression of different pathological states. Jaganjac have described the role of HNE as second messengers of free radicals in the pathogenesis of diabetes mellitus (DM) (Jaganjac et al. 2013). Chapple summarized the production, metabolism and consequences of HNE synthesis within vascular endothelial, smooth muscle cells and targeted signaling within vasculature (Chapple et al. 2013). Review focuses on the role of HNE affecting cell signaling pathways and endothelial barrier dysfunction through modulation of the activities of proteins/enzymes by Michael adducts formation, enhancing the level of protein tyrosine phosphorylation of the target proteins, and by reorganization of cytoskeletal, focal adhesion, and adherens junction proteins (Usatyuk and Natarajan 2012). Perluigi summarized the role of HNE-induced protein modification in neurodegenerative diseases (Perluigi et al. 2012).

Although lipid peroxidation generates far more HNE than ACR (Stevens and Maier 2008), it has been reported, based on the urinary daily excretion, that production of ACR metabolites in humans is 102-fold greater than that of HNE metabolites (Stevens and Maier 2008). This clearly indicates that sources other than lipid peroxidation, such as amino acids and polyamines, are mainly involved in ACR production. ACR reacts preferentially with Cys, Lys, and His residues (the Lys adducts being the more stable products) via Michael-type addition reactions preserving the aldehyde functionality on the modified protein, rather than by Schiff's base formation. The order of ACR reactivity for protein nucleophiles (Cys>His>Lys) is the same as that established for HNE (Aldini et al. 2007; Carini et al. 2004; Schaur 2003). The reaction of ACR with Lys may result in β-substituted propanals (R-NH-CH_2-CH_2-CHO), but the major adduct formed on reaction with protein is the Nε-(3-formyl-3,4-dehydropiperidino)lysine adduct (FDP-Lys) (Uchida et al. 1998). This compound is a reactive intermediate that can covalently bind to thiols, such as glutathione (GSH) (Furuhata et al. 2002), through the retained electrophilic carbonyl moiety. Another ACR-Lys adduct, Nε-(3-methylpyridinium)lysine (MP-Lys), was identified later and found to be a highly stable end product (Furuhata et al. 2003).

Major pathways dysregulated by ACR and related to the cytotoxic effect include those involved in apoptosis, cell cycle control, transcription, cell signaling, and protein biosynthesis (Thompson and Burcham 2008). LoPachin et al. (2009) reviewed the adduct chemistry of the α,β-unsaturated carbonyls and possible protein targets, and suggested that protein adduct chemistry of the conjugated type-2 alkenes might provide insight into the pathogenesis of certain human neurodegenerative diseases. Type-2 alkenes share in fact a common mechanism of action at nerve terminals in the brain, and the combined effects of these substances could contribute to some neurologic disorders (LoPachin et al. 2009). In a very recent review focused on the ACR involvement in neurodisorders (Weinhold 2011), the authors pointed out that one of the emerging suspected culprits in their development is ACR, which tends to be significantly elevated in the brains or spinal cords of people who have Alzheimer's disease, PD, amyotrophic lateral sclerosis, and other neurologic disorders (LoPachin et al. 2009; Singh et al. 2010; Tanuma et al. 2008; Jomova et al. 2010; Nam et al. 2010). ACR (and protein-conjugated ACR), by accumulating in the blood of patients with chronic renal failure, may function as uremic toxin which accelerates the progression of uremia (Igarashi et al. 2006). Both GLX and MGLX are capable of causing cellular damage and apoptosis (Wolff et al. 1991). They are involved in the formation of advanced glycation end-products (AGEs) which have been linked to long-term sequela of chronic diseases such as diabetic retinopathy, neuropathy, and nephropathy (Abordo et al. 1999). Moreover, GLX and MGLX can form adducts with nucleic acids bases inducing severe genotoxicity. GLX and MGLX are primarily detoxified by the glyoxalase system present in the cells of bacteria, protozoa, fungi, plants, animals, and humans (Thornalley 1993), whose functionality is strictly related to the bioavailability of its cofactor, GSH; this is the reason why oxidative stress conditions involving a dramatic decrease of cellular levels of GSH are typically associated with an overproduction of GLX and MGLX.

Oxidative Stress-Related Aldehydes in Disease States: An Overall View

Oxidative stress induced by α,β-unsaturated aldehydes plays an important role in the pathogenesis of many diseases through direct injury to the endothelium, as well as through redox-sensitive mechanisms (Table 2). α,β-Unsaturated aldehydes increase oxidative stress in endothelial, macrophage, and smooth muscle cells which in turn induces a proinflammatory vascular phenotype by stimulating the α,β-Unsaturated aldehydes tend to trigger the formation of ROS or act as oxidants and potentiate oxidative stress in cells (Lee et al. 2011).

Table 2. OSRAs and other biomarkers in oxidative stress-related diseases (Dalle-Donne et al. 2006; Lee and Park 2013)

		Disease									
		Alzheimer D.	Atherosclerosis	Diabete Mellitus	Cardiovascular Diseases	Pulmonary Diseases	Parkinson's Disease	Inflammation	Fanconi's Anaemia	Cancer	Down Syndrome
OSRAs	ACR	↑	↑	↑	↑	↑				↑	
	CRA	↑				↑					
	MDA	↑	↑	↑	↑		↑	↑	↑	↑	
	HNE	↑	↑	↑	↑	↑	↑	↑		↑	
	GLX		↑	↑				↑			
	MGLX		↑	↑				↑	↑		
Oxidative stress parameters	GSH/GSSG	↓		↓	↓	↓			↓	↓	↓
	F_2-Isoprostane	↑	↑	↑	↑	↑					↑
	Tyr-NO_2	↑	↑	↑		↑		↑			
	Carbonylated Proteins	↑		↑		↑	↑				

Adams Jr. and Klaidman reported that ACR was oxidized by xanthine oxidase to produce oxygen radicals and that the GSH adduct of ACR also induced oxygen radical formation (Adams and Klaidman 1993). ACR depletes endogenous GSH which itself is a critical component of the endogenous antioxidant defense system, thereby increasing the ROS levels (Kehrer and Biswal 2000).

In addition, it has been shown that HNE mediates endothelial nitric oxide synthase (eNOS) uncoupling and superoxide generation by altering tetrahydrobiopterin homeostasis (Whitsett et al. 2007) and that it induces ROS generation by activating nicotinamide adenine dinucleotide phosphate (NADPH) oxidase which is dependent on the activity of 5-lipoxygenase (5-LO) (Yun et al. 2010).

α,β-Unsaturated Aldehydes and Antioxidant Enzymes

Oxidative (electrophilic) stress induces NF-E2-related factor 2 (Nrf2)/antioxidant response element-mediated expression of phase II detoxifying and antioxidant enzymes and activates other stress-inducible genes (Nguyen et al. 2004). α,β-Unsaturated aldehydes are attracted to electrons and can inactivate the nucleophilic active sites of thiolate or selenocysteine enzymes, such as glutathione peroxidase (GPx) through covalent bonding. The inactivation of GPx by α,β-unsaturated aldehydes is involved in imbalance of the redox state in cell (Miyamoto et al. 2003). The thioredoxin (Trx)/thioredoxin reductase (TR) system plays a crucial role in many biological functions, such as redox regulation, apoptosis, and immunomodulation in diverse organisms. Endothelial cells exposed to ACR show rapidly inactivation of TR, resulting in an increase in oxidative cellular damage. In ACR-stimulated human umbilical vein endothelial cells (HUVECs), the induction of heat shock protein 72 (Hsp72) is considered to be a defense system unique to HUVECs (Misonou et al. 2005). The results of some studies indicate that a highly electrophilic compound, such as ACR, would have the potential to increase Nrf2-mediated gene expression, including that of the cytoprotective antioxidant heme oxygenase-1 (HO-1) in macrophages (Lee et al. 2009) and endothelial cells (Wu et al. 2006). Furthermore, HNE and CRA induces HO-1 expression in endothelial cells (Ishikado et al. 2010). HO-1, a rate-limiting enzyme in heme metabolism, has been recognized as an important factor that protects vascular tissue against atherosclerosis by exerting antioxidative, anti-inflammatory, antiproliferative, anti-apoptotic, and vasodilatory effects on the vasculature. Therefore, increased HO-1 expression in various cells treated with α,β-unsaturated aldehydes may serve as an adaptive response to oxidative damage.

α,β-Unsaturated Aldehydes in the Pathogenesis of Vascular Diseases

Vascular disease is a complex inflammatory disease that involves several types of inflammatory cells, multiple inflammatory mediators, and oxidative stress. α,β-Unsaturated aldehydes cause inflammation and damage cells by inducing oxidative stress, redox-sensitive mechanisms, and proinflammatory mediators. The results of many studies have implicated α,β-unsaturated aldehydes in the pathogenesis of vascular disease.

Maintaining the redox balance in the vascular system is of paramount importance since uncompensated oxidative stress contributes to endothelial dysfunction and vascular disease. Oxidative stress is increasingly seen as a major upstream component in the signaling cascade involved in many cellular functions, such as cell proliferation, inflammatory responses, adhesion molecule stimulation, and chemoattractant production. The mechanisms by which endothelial oxidative stress leads to vascular inflammation and the development of atherosclerosis have been reported (Foncea et al. 2000).

α,β-Unsaturated Aldehydes and Inflammation

α,β-Unsaturated aldehyde-induced toxicity is reported to occur because of depletion of cellular GSH, which subsequently induces ROS production that leads to cell malfunction. ROS was also shown to induce the production of various atherogenic factors, including inflammatory mediators.

α,β-Unsaturated Aldehydes and Nuclear Factor Kappa-Light-Chain-Enhancer of Activated B Cells

The nuclear factor kappa-light-chain-enhancer of activated B cells (NF-κB)/Rel family complex is a redox-sensitive transcription factor that plays a role in the expression of various rapid-response genes associated with the inflammatory and immune responses. In addition, NF-κB activation may play a role in the development of chronic inflammatory diseases, such as rheumatoid arthritis, Alzheimer's disease, and atherosclerosis.

The results of many studies suggest that α,β-unsaturated aldehydes can regulate inflammation by modulating NF-κB signaling (Kehrer and Biswal 2000). ACR may affect NF-κB activation, either indirectly by decreasing cellular reduced GSH content or directly by binding to the reactive cysteine on the subunit of IκB kinase (IKK) (Kehrer and Biswal 2000). The effect of ACR on NF-κB may be cell-type specific and other regulatory mechanisms may also be involved. Li et al. (1999) reported that ACR induced IκB expression in rat alveolar macrophage cells, an effect that led to the inhibition of NF-κB activation. However, Haberzettl et al. (2009) showed that the ACR-induced increase in cytokine production was accompanied by NF-κB activation (Niedernhofer et al. 2003). The other α,β-unsaturated aldehyde, HNE, may also play a role in modulating NF-κB activation through a mechanism similar to that of ACR. It has been suggested that HNE induces 5-LO expression via epidermal growth factor receptor (EGFR)-mediated activation of the NF-κB/extracellular-regulated kinase (ERK) pathways in murine macrophages (Lee et al. 2010).

α,β-Unsaturated Aldehydes and Proinflammatory Mediators

Cyclooxygenase (COX)-2 is an inducible isoform of COX, which is the key enzyme that regulates the amount of and the duration for which proinflammatory prostaglandins (PG) are produced and also plays a crucial role in inflammation. Under normal conditions, COX-2 expression is tightly regulated, but it is dramatically induced during inflammation by various stimulants. Burleigh et al. suggested that COX-2 expression promotes atherosclerotic inflammation (Burleigh et al. 2002). Since chronic inflammation plays a significant role in atherosclerosis, COX-2 may participate in the development of atherosclerosis.

The endothelium is a vulnerable target for ACR and related aldehydes. Several studies have reported that exposure to ACR causes endothelial damage. Endothelial cells exposed to ACR exhibit a time-and dose-dependent stimulation of COX-2 expression and enhancement of PG synthesis. The increased PG synthesis in endothelial cells after treatment with ACR reflects an increase in the levels of functional COX-2 protein. In addition, the induction of COX-2 by ACR occurs through activation of the protein kinase C (PKC), p38 mitogen-activated protein kinase (MAPK), and cAMP response element-binding protein (CREB) pathways; it has been suggested that ACR plays an important role in the progression of atherosclerosis via an inflammatory response involving COX-2 expression. HNE is reported to strongly induce COX-2 expression in macrophages. These data suggest that the HNE accumulated in macrophages/foam cells functions as an inflammatory mediator that plays a role in stimulating the inflammatory response and contributes to the progression of atherogenesis.

In addition, Haberzettl showed that ACR treatment increased the production of interleukin-6 (IL-6), tumor necrosis factor-α (TNF-α), and interleukin-8 (IL-8) in endothelial cells (Haberzettl et al. 2009). These findings suggest new proinflammatory and atherogenic aspects of ACR toxicity and the possibility that endogenously produced ACR can contribute toward endothelial injury and inflammation. Because the induction of cytokines, such as TNF-α, IL-6, and IL-8, plays a crucial role in atherosclerosis, production of these cytokines may be a significant feature of atherogenesis. Furthermore, ACR treatment induced endoplasmic reticulum (ER) stress and triggered the unfolded protein response (Haberzettl et al. 2009).

Activated macrophages are reported to generate and secrete matrix metalloproteinase (MMP)-9 which degrades atherosclerotic plaque constituents. A recent study by O'Toole et al. (2009) reported that secretion of MMP-9 increases in ACR-stimulated human macrophages. In addition, murine macrophages exposed to ACR exhibited 5-LO overexpression, subsequent proinflammatory leukotriene (LT) accumulation, and enhanced MMP-9 biosynthesis (Kim et al. 2010). These data support the possibility that exposure to oxidants or acute inflammatory events can trigger plaque rupture. Akiba et al. (2006) showed that HNE accelerates MMP-1 production in human coronary smooth muscle cells (hCSMCs). MMP-1 (collagenases) cleave native collagen types I and III, which are predominant structural components of atherosclerotic lesions, indicating that increase in the levels of collagenases is a critical event in the progression of atherosclerosis.

Oxidative Stress-Related Aldehydes in Fanconi Anemia

Fanconi Anemia (FA) is a severe hereditary disorder with progressive bone marrow failure (BMF), developmental abnormalities, and increased susceptibility to both haematological (typically myelodysplastic syndrome and acute myeloid leukemia) and non-hematological malignancies. Major aspects of the FA's clinical phenotype include short stature, skeletal abnormalities (most frequently of the thumbs or limbs), and abnormal skin pigmentation, but variable developmental abnormalities of other organs also occur (Auerbach et al. 1998; Alter 1996; Ahmad et al.2002). At present, there are two alternative theories trying to address the pathogenesis of FA, one focused on the defect of FA cells to repair DNA damage induced by "crosslinkers", i.e., agents causing DNA interstrand crosslinks (reviewed by Kupfer 2013), and the other theory focused on cell inability to handle oxidative stress (reviewed by Pagano et al. 2013).

It was assumed that FA's crosslinker sensitivity was due to a deficiency in DNA repair pathways since early reports dating back to mid 1970s (Auerbach and Wolman 1976), in analogy with other genetic diseases categorized for DNA repair defects. But since the pioneering report by Ingrid Emerit (1977), who found SOD-induced decrease of FA cell chromosomal instability, which was confirmed in MMC-treated FA cells, a series of independent investigations related FA phenotype with oxidative stress.

Excess oxygen and iron sensitivity was reported in FA cells, both involving cell viability and G2 cell growth phase arrest (Joenje et al. 1981; Poot et al. 1996).

Excess oxidative DNA damage was observed in FA lymphoblasts, which was enhanced in MMC-exposed cells, and was partly attributed by Takeuchi and Morimoto (1993) to decreased CAT activity, while Trx overexpression counteracted MMC- and DEB-induced toxicity in FA fibroblasts (Ruppitsch et al. 1998).

Unconfined to *in vitro* testing, a series of independent studies conducted on blood cells and biological fluids from FA patients found an *in vivo* prooxidant state assessed on a set of redox endpoints, including oxidative DNA damage, GSH oxidation, WBC chemiluminescence, SOD and CAT expression, and excess aldehyde levels, including methylglyoxal and malondialdehyde (Degan et al. 1995; Pagano et al. 2004; Petrović et al. 2011).

As previously discussed, besides being a footprint of ROS generation, aldehydes can trigger a cascade of toxicity pathways which may ultimately lead to citotoxicity, by protein function impairment, and genotoxicity, by DNA cross-linkings (Figure 3). The key role played by OSRAs has been recently supported by a series of reports focusing on a deficiency of aldehyde dehydrogenase 2 (ALDH2) in FANCD2 cells and fancd2 (-/-) mice (Marietta et al. 2009; Langevin et al. 2011; Garaycoechea et al. 2012), and relating this deficiency to inability to counteract acetaldehyde toxicity. The relevance of these data has found support by the report of Hira (2013), who found that a ALDH2 variant is associated with accelerated progression of bone marrow failure in Japanese FA patients.

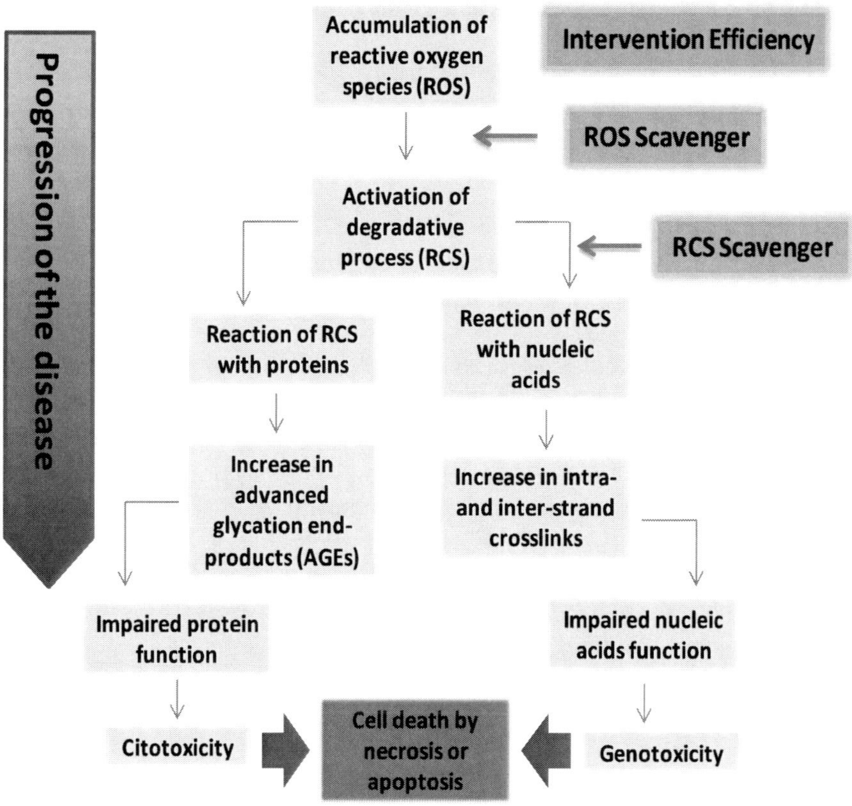

Figure 3. Cascade of events triggered by oxidative stress and mediated by OSRAs.

Conclusion

Overwhelming evidence during the last 10 years indicates the involvement of protein, enzyme and nucleic acid modification by reactive α,β-unsaturated aldehydes in different pathological conditions typically associated with advanced OS, e.g., atherosclerosis, aging, inflammation, diabetes, neurodegenerative diseases, and a number of genetic diseases including FA.

Also documented in the literature, the complex cascade of events leading to an uncontrolled OS is the main cause of cells and tissue injury. The over-production of ROS triggers a series of cyto- and genotoxic processes leading to cell dysfunction and/or transformation and/or death. The primary sources of endogenous ROS production are the mitochondria, plasma membrane, endoplasmic reticulum, and peroxisomes through a variety of mechanisms including enzymatic reactions and/or auto-oxidation of several compounds. On the other hand, excess ROS production results in broad array of oxidative damage of different biomolecules (nucleic acids, proteins, lipids and carbohydrates). Thus OSRAs represent just one of several by-products of an OS condition.

Thus, any attempt to specifically counteract OSRA toxicity might fail to target the underlying causes of a pro-oxidant state a detected in FA. The present knowledge warrants more comprehensive interventions, relying on evidence-based design of animal and clinical studies.

References

Abordo, E. A., Minhas, H. S. & Thornalley, P. J. (1999). Accumulation of α-oxoaldehydes during oxidative stress: a role in cytotoxicity. *Biochem. Pharmacol., 58*, 641–648.

Adams Jr., J. D. & Klaidman, L. K. (1993). Acrolein-induced oxygen radical formation. *Free Radic. Biol. Med., 15*, 187–193.

Ahmad, S. I., Hanaoka, F. & Kirk, S. H. (2002). Molecular biology of Fanconi anaemia: an old problem, a new insight. *Bioessays, 24*, 439-448.

Akiba, S., Kumazawa, S., Yamaguchi, H., Hontani, N., Matsumoto, T., Ikeda, T., Oka, M. & Sato, T. (2006). Acceleration of matrix metalloproteinase-1 production and activation of platelet-derived growth factor receptor beta in human coronary smooth muscle cells by oxidized LDL and 4-hydroxynonenal. *Biochim. Biophys. Acta, 1763*, 797–804.

Alary, J., Guéraud, F. & Cravedi, J.-P. (2003). Fate of 4-hydroxynonenal in vivo: disposition and metabolic pathways. *Mol. Asp. Med., 24*, 177–187.

Aldini, G., Dalle-Donne, I., Maffei Facino, R., Milzani, A. & Carini, M. (2007). Intervention strategies to inhibit protein carbonylation by lipoxidation-derived reactive carbonyls. *Med. Res. Rev., 27*, 817–868.

Alter, B. P. (1996). Fanconi's anemia and malignancies. *Am. J. Hematol., 53*, 99-110.

Anderson, M. M., Hazen, S. L., Hsu, F. F. & Heinecke, J. W. (1997). Human neutrophils employ the myeloperoxidase-hydrogen peroxide-chloride system to convert hydroxy-amino acids into glycolaldehyde, 2-hydroxypropanal, and acrolein. A mechanism for the generation of highly reactive alpha-hydroxy and alpha,beta-unsaturated aldehydes by phagocytes at sites of inflammation. *J. Clin. Invest., 99*, 424–432.

Auerbach, A. D. & Wolman, S. R. (1976). Susceptibility of Fanconi's anaemia fibroblasts to chromosome damage by carcinogens. *Nature, 261*, 494-496.

Auerbach, A. D., Buchwald, M. & Joenje, H. (1998). Fanconi anemia. In: *The Genetic Basis of Human Cancer*. Vogelstein, B. and Kinzier, K.W., eds., McGraw-Hill, New York, NY, 317-322.

Ayala, A., Muñoz, M. F. & Arguelles, S. (2014). Lipid peroxidation: production, metabolism, and signaling mechanisms of malondialdehyde and 4-hydroxy-2-nonenal. *Oxid. Med. Cell. Longev.*, 2014, 360438.

Balogh L. M. & Atkins W. M. (2011). Interactions of glutathione transferases with 4-hydroxynonenal. *Drug Met. Rev., 43*, 165–178.

Batista, C. K., Mota, J. M., Souza, M. L., Leitao, B. T., Souza, M. H., Brito, G. A., Cunha, F. Q. & Ribeiro, R. A. (2007). Amifostine and glutathione prevent ifosfamide- and acrolein-induced hemorrhagic cystitis. *Cancer Chemother. Pharmacol., 59*, 71–77.

Burleigh, M. E., Babaev, V. R., Oates, J. A., Harris, R. C., Gautam, S., Riendeau, D., Marnett, L. J., Morrow, J. D., Fazio, S. & Linton, M. F. (2002). Cyclooxygenase-2 promotes early atherosclerotic lesion formation in LDL receptor-deficient mice. *Circulation, 105*, 1816–1823.

Butterfield, D. A., Bader Lange, M. L. & Sultana, R. (2010). Involvements of the lipid peroxidation product, HNE, in the pathogenesis and progression of Alzheimer's disease. *Biochim. Biophys. Acta, 1801*, 924–929.

Carini, M., Aldini, G. & Maffei Facino, R. (2004). Mass spectrometry for detection of 4-hydroxy-trans-2-nonenal (HNE) adducts with peptides and proteins. *Mass Spectrom. Rev., 23*, 281–305.

Casazza, J. P., Felver, M. E. & Veech, R. L. (1984). The metabolism of acetone in rat. *J. Biol. Chem., 259*, 231-236.

Chapple, S. J., Cheng, X. & Mann, G. E. (2013). Effects of 4-hydroxynonenal on vascular endothelial and smooth muscle cell redox signaling and function in health and disease. *Redox Biol., 1*, 319–331.

Cheng, J., Wang, F., Yu, D.-F., Wu, P.-F. & Chen, J.-G. (2011). The cytotoxic mechanism of malondialdehyde and protective effect of carnosine via protein cross-linking/mitochondrial dysfunction/reactive oxygen species/MAPK pathway in neurons. *Eur. J. Pharmacol., 650*, 184–194.

Conklin, D. J., Haberzettl, P., Lesgards, J. F., Prough, R. A., Srivastava, S. & Bhatnagar, A. (2009). Increased sensitivity of glutathione S-transferase P-null mice to cyclophosphamide-induced urinary bladder toxicity. *J. Pharmacol. Exp. Ther., 331*, 456–469.

Dalleau, S., Baradat, M., Guéraud, F. & Huc, L. (2013). Cell death and diseases related to oxidative stress: 4-hydroxynonenal (HNE) in the balance. *Cell Death Diff., 20*, 1615–1630.

Dalle-Donne, I., Rossi, R., Colombo, R., Giustarini, D. & Milzani, A. (2006). Biomarkers of oxidative damage in human disease. *Clin. Chem., 52*, 601–623.

Degan, P., Bonassi, S., De Caterina, M., Korkina, L. G., Pinto, L., Scopacasa, F., Zatterale, A., Calzone, R. & Pagano, G. (1995). In vivo accumulation of 8-hydroxy-2'-deoxyguanosine in DNA correlates with release of reactive oxygen species in Fanconi's anaemia families. *Carcinogenesis, 16*, 735-742.

Del Rio, D., Stewart, A. J. & Pellegrini, N. (2005). A review of recent studies on malondialdehyde as toxic molecule and biological marker of oxidative stress. *Nutr. Met. Cardiovas. Dis., 15*, 316-328.

Eder, E., Schuler, D. & Budiawan, B. (1999). Cancer risk assessment for crotonaldehyde and 2-hexenal: an approach. *IARC Scient. Publ., 150*, 219–232.

Eder, E. & Budiawan, B. (2001). Cancer risk assessment for the environmental mutagen and carcinogen crotonaldehyde on the basis of TD50 and comparison with 1,N2-propanodeoxyguanosine adduct levels. *Cancer Epidemiol. Biomarkers Prev., 10*, 883–888.

Einhorn, I. N. (1975). Physiological and toxicological aspects of smoke produced during the combustion of polymeric materials. *Environ. Health Perspect., 11*, 163–189.

Esterbauer, H., Schaur, R .J. & Zollner, H. (1991). Chemistry and Biochemistry of 4-hydroxynonenal,malonaldehyde and related aldehydes. *Free Radic. Biol. Med., 11*, 81–128.

Faroon, O., Roney, N., Taylor, J., Ashizawa, A., Lumpkin, M. H. & Plewak, D. J. (2008). Acrolein health effects. *Toxicol. Ind. Health,. 24*, 447-490.

Feron, V. J., Til, H. P., de Vrijer, F., Woutersen, R. A., Cassee, F. R. & van Bladeren, P. J. (1991). Aldehydes: occurrence, carcinogenic potential, mechanism of action and risk assessment. *Mutat. Res., 259*, 363–385.

Foncea, R., Carvajal, C., Almarza, C. & Leighton, F. (2000). Endothelial cell oxidative stress and signal transduction. *Biol. Res., 33*, 89–96.

Frankel, E. N. & Neff, W. E. (1983). Formation of malonaldehyde from lipid oxidation products. *Biochim. Biophys. Acta, 754*, 264-270.

Fullana, A., Carbonell-Barrachina, A. A. & Sidhu, S. (2004). Comparison of volatile aldehydes present in the cooking fumes of extra virgin olive, olive, and canola oils. *J. Agric. Food Chem., 52*, 5207–5214.

Furuhata, A., Nakamura, M., Osawa, T. & Uchida, K. (2002). Thiolation of protein-bound carcinogenic aldehyde. An electrophilic acrolein-lysine adduct that covalently binds to thiols. *J. Biol. Chem., 277*, 27919–27926.

Furuhata, A., Ishii, T., Kumazawa, S., Yamada, T., Nakayama, T. & Uchida, K. (2003). Nε-(3-methylpyridinium)lysine, a major antigenic adduct generated in acrolein-modified protein. *J. Biol. Chem., 278*, 48658–48665.

Garaycoechea, J. I., Crossan, G. P., Langevin, F., Daly, M., Arends, M. J. & Patel, K. J. (2012). Genotoxic consequences of endogenous aldehydes on mouse haematopoietic stem cell function. *Nature, 489*, 571-575.

Haberzettl, P., Vladykovskaya, E., Srivastava, S. & Bhatnagar, A. (2009). Role of endoplasmic reticulum stress in acrolein-induced endothelial activation. *Toxicol. Appl. Pharmacol., 234*, 14–24.

Hebert, G. B., Deroussent, G. & Veal, A. (2010). Oxazaphosphorines: new therapeutic strategies for an old class of drugs. *Expert Opin. Drug Metab. Toxicol., 6*, 919–938.

Hecker, M. & Ullrich, V. (1989). On the mechanism of prostacyclin and thromboxane A2 biosynthesis. *J. Biol. Chem., 264*, 141-150.

Hira, A., Yabe, H., Yoshida, K., Okuno, Y., Shiraishi, Y., Chiba, K., Tanaka, H., Miyano, S., Nakamura, J., Kojima, S., Ogawa, S., Matsuo, K., Takata, M. & Yabe, M. (2013). Variant ALDH2 is associated with accelerated progression of bone marrow failure in Japanese Fanconi anemia patients. *Blood, 122*, 3206-3209.

Igarashi, K., Ueda, S., Yoshida, K. & Kashiwagi, K. (2006). Polyamines in renal failure. *Amino Acids, 31*, 477–483.

Ishikado, A., Nishio, Y., Morino, K., Ugi, S, Kondo, H., Makino, T., Kashiwagi, A. & Maegawa, H. (2010). Low concentration of 4-hydroxy hexenal increases heme oxygenase-1 expression through activation ofNrf2 and antioxidative activity in vascular endothelial cells. *Biochem. Biophys. Res. Commun., 402*, 99–104.

Jacobs, A. T. & Marnett, L. J. (2010). Systems analysis of protein modification and cellular responses induced by electrophile stress. *Acc. Chem. Res., 43*, 673–683.

Jaganjac, M., Tirosh, O., Cohen, G., Sasson, S. & Zarkovic, N. (2013). Reactive aldehydes - second messengers of free radicals in diabetes mellitus. *Free Radic. Res., 47*, 39–48.

Ji, C., Rouzer, C. A., Marnett, L. J. & Pietenpol, J. A. (1998). Induction of cell cycle arrest by the endogenous product of lipid peroxidation, malondialdehyde. *Carcinogenesis, 19*, 1275–1283.

Joenje, H., Arwert, F., Eriksson, A. W., de Koning, H. & Oostra, A. B. (1981). Oxygen-dependence of chromosomal aberrations in Fanconi's anaemia. *Nature, 290*, 142-143.

Jomova, K., Vondrakova, D., Lawson, M. & Valko, M. (2010). Metals, oxidative stress and neurodegenerative disorders. *Mol. Cell. Biochem., 345*, 91–104.

Kawaguchi-Niida, M., Shibata, N., Morikawa, S., Uchida, K., Yamamoto, T., Sawada, T. & Kobayashi, M. (2006). Crotonaldehyde accumulates in glial cells of Alzheimer's disease brain. *Acta Neuropathologica, 111*, 422–429.

Kehrer, J. P. & Biswal, S. S. (2000). The molecular effects of acrolein. *Toxicol. Sci., 57*, 6–15.

Kielhorn, J., Pohlenz-Michel, C. & Schmidt, S. (2004). *Glyoxal. Concise International Chemical Assessment Document 57*, World Health Organization, Geneva, Switzerland.

Kim, C. E., Lee, S. J., Seo, K. W., Park, K. M., Yun, J. W., Bae, J. U., Bae, S. S. & Kim, C. D. (2010). Acrolein increases 5-lipoxygenase expression in murine macrophages through activation of ERK pathway. *Toxicol. Appl. Pharmacol., 245*, 76–82.

Kode, A., Yang, S. & Rahman, I. (2006). Differential effects of cigarette smoke on oxidative stress and proinflammatory cytokine release in primary human airway epithelial cells and in a variety of transformed alveolar epithelial cells. *Resp. Res., 7*, 132.

Korkmaz, A., Topal, T. & Oter, S. (2007). Pathophysiological aspects of cyclophosphamide and ifosfamide induced hemorrhagic cystitis; implication of reactive oxygen and nitrogen species as well as PARP activation. *Cell. Biol. Toxicol.*, 23, 303–312.

Kumagai, T., Matsukawa, N., Kaneko, Y., Kusumi, Y., Mitsumata, M. & Uchida, K. (2004). A lipid peroxidation-derived inflammatory mediator: identification of 4-hydroxy-2-nonenal as a potential inducer of cyclooxygenase-2 in macrophages. *J. Biol. Chem., 279*, 48389–48396.

Kupfer, G. M. (2013). Fanconi anemia: a signal transduction and DNA repair pathway. *Yale J. Biol. Med., 86*, 491-497.

Lange, J. N., Wood, K. D., Knight, J., Assimos, D. G. & Holmes, R. P. (2012). Glyoxal Formation and Its Role in Endogenous Oxalate Synthesis. *Adv. Urology,* 2012, 819202.

Langevin, F., Crossan, G. P., Rosado, I. V., Arends, M. J. & Patel, K. J. (2011). Fancd2 counteracts the toxic effects of naturally produced aldehydes in mice. *Nature, 475*, 53-58.

Lee, N. J., Lee, S. E., Lee, S. H., Ryu, D. S. & Park, Y. S. (2009). Acrolein induces adaptive response through upregulate ofHO-1 via activation of Nrf2 in RAW264. 7 macrophage. *Mol. Cell. Toxicol., 5*, 77.

Lee, S. E., Jeong, S. I., Kim, G. D., Yang, H., Park, C. S., Jin, Y. H. & Park, Y. S. (2011). Upregulation of heme oxygenase-1 as an adaptive mechanism for protection against crotonaldehyde in human umbilical vein endothelial cells. *Toxicol. Lett., 201*, 240–248.

Lee, S. E. & Park, Y. S. (2013). Role of lipid peroxidation-derived α, β-unsaturated aldehydes in vascular dysfunction. *Oxid. Med. Cell. Longev.*, 2013, 629028.

Lee, S. J., Kim, C. E., Seo, K. W. & Kim, C. D. (2010). HNE-induced 5-LO expression is regulated by NF-κB/ERK and Sp1/p38 MAPK pathways via EGF receptor in murine macrophages. *Cardiovasc. Res., 88*, 352–359.

Li, L., Hamilton Jr., R. F. & Holian, A. (1999). Effect of acrolein on human alveolar macrophage NF-κB activity. *Am. J. Physiol., 277*, L550–L557.

Liu, X., Yang, Z., Pan, X., Zhu, M. & Xie, J. (2010). Crotonaldehyde induces oxidative stress and caspase-dependent apoptosis in human bronchial epithelial cells. *Toxicol. Lett., 195*, 90–98.

Loeb, L. A., James, E. A., Waltersdorph, A. M. & Klebanoff, S. J. (1988). Mutagenesis by the autoxidation of iron with isolated DNA. *Proc. Natl. Acad. Sci. USA, 85*, 3918–3922.

Loidl-Stahlhofen, A. & Spiteller, G. (1994). α-Hydroxyaldehydes, products of lipid peroxidation. *Biochim. Biophys. Acta, 1211*, 156–160.

LoPachin, R. M., Gavin, T. & Barber, D. S. (2008). Type-2 alkenes mediate synaptotoxicity in neurodegenerative diseases. *Neurotoxicology, 29*, 871–882.

LoPachin, R. M., Gavin, T., Petersen, D. R. & Barber, D. S. (2009). Molecular mechanisms of 4-hydroxy-2-nonenal and acrolein toxicity: nucleophilic targets and adduct formation. *Chem. Res. Toxicol., 22*, 1499–1508.

Manini, P., La Pietra, P., Panzella, L., Napolitano, A. & d'Ischia, M. (2006). Glyoxal formation by Fenton-induced degradation of carbohydrates and related compounds. *Carbohydr. Res., 341*, 1828–1833.

Marietta, C., Thompson, L. H., Lamerdin, J. E. & Brooks, P. J. (2009). Acetaldehyde 474 stimulates FANCD2 monoubiquitination, H2AX phosphorylation, and BRCA1 phosphorylation in human cells in vitro: implications for alcohol-related carcinogenesis. *Mutat. Res., 664*, 77-83.

Marnett, L. J. (1999). Lipid peroxidation-DNA damage by malondialdehyde. *Mut. Res., 424*, 83–95.

Marnett, L. J. (2002). Oxy radicals, lipid peroxidation and DNA damage. *Toxicology*, 181-182, 219–222.

Marnett, L. J., Riggins, J. N. & West, J. D. (2003). Endogenous generation of reactive oxidants and electrophiles and their reactions with DNA and protein. *J. Clinical Invest., 111*, 583–593.

Misonou, Y., Takahashi, M., Park, Y. S., Asahi, M., Miyamoto, Y., Sakiyama, H., Cheng, X. & Taniguchi, N. (2005). Acrolein induces Hsp72 via both PKCδ/JNK and calcium signaling pathways in human umbilical vein endothelial cells. *Free Radic. Res., 39*, 507–512.

Miyamoto, Y., Koh, Y. H., Park, Y. S., Fujiwara, N., Sakiyama, H., Misonou, Y., Ookawara, T., Suzuki, K., Honke, K. & Taniguchi, N. (2003). Oxidative stress caused by inactivation of glutathione peroxidase and adaptive responses. *Biol. Chem., 384*, 567–574.

Murata-Kamiya, N., Kamiya, H., Iwamoto, N. & Kasai, H. (1995). Formation of a mutagen, glyoxal, from DNA treated with oxygen free radicals. *Carcinogenesis, 16*, 2251–2253.

Nam, D. T., Arseneault, M., Murthy, V. & Ramassamy, C. (2010). Potential role of acrolein in neurodegeneration and in Alzheimer's disease. *Curr. Mol. Pharmacol., 3*, 66–78.

Napoli, C., D'Armiento, F. P., Mancini, F.P., Postiglione, A., Witztum, J. L., Palumbo, G. & Palinski, W. (1997). Fatty streak formation occurs in human fetal aortas and is greatly enhanced maternal, hypercholesterolemia. Intimal accumulation of low density lipoprotein and its oxidation precede monocyte recruitment into early atheroeclerotic lesions. *J. Clin. Invest., 100*, 2680–2690.

Negre-Salvayre, A., Coatrieux, C., Ingueneau, C. & Salvayre, R. (2008). Advanced lipid peroxidation end products in oxidative damage to proteins. Potential role in diseases and therapeutic prospects for the inhibitors. *Brit. J. Pharmacol., 153*, 6–20.

Neudecker, T., Eder, E., Deininger, C. & Henschler, D. (1989). Crotonaldehyde is mutagenic in Salmonella typhimurium TA100. *Environ. Mol. Mutagen., 14*, 146–148.

Nguyen, T., Yang, C. S. & Pickett, C. B. (2004). The pathways and molecular mechanisms regulating Nrf2 activation in response to chemical stress. *Free Radic. Biol. Med., 37*, 433–441.

Niedernhofer, L. J., Daniels, J. S., Rouzer, C. A., Greene, R. E. & Marnett, L. J. (2003). Malondialdehyde, a product of lipid peroxidation, is mutagenic in human cells. *J. Biol. Chem., 278*, 31426–31433.

Nordenson, I. (1977). Effect of superoxide dismutase and catalase on spontaneously occurring chromosome breaks in patients with Fanconi's anemia. *Hereditas, 86*, 147-150.

O'Toole, T. E., Zheng, Y., Hellmann, J., Conklin, D. J., Barski, O. & Bhatnagar, A. (2009). Acrolein activatesmatrixmetalloproteinases by increasing reactive oxygen species in macrophages. *Toxicol. Appl. Pharmacol., 236*, 194–201.

Pagano, G., Degan, P., d'Ischia, M., Kelly, F. J., Pallardo, F. V., Zatterale, A., Anak, S. S., Akisik, E. E., Beneduce, G., Calzone, R., De Nicola, E., Dunster, C., Lloret, A., Manini, P., Nobili, B., Saviano, A., Vuttariello, E. & Warnau, M. (2004). Gender- and age-related distinctions for the in vivo prooxidant state in Fanconi anaemia patients. *Carcinogenesis, 25*, 1899-1909.

Pagano, G, Aiello Talamanca, A, Castello, G, d'Ischia, M., Pallardó, F. V., Petrović, S., Porto, B., Tiano, L. & Zatterale, A. (2013). From clinical description to in vitro and animal studies, and backwards to patients: Oxidative stress and mitochondrial dysfunction in Fanconi anemia. *Free Radic. Biol. Med., 58*, 118-125.

Park, Y. S., Misonou, Y., Fujiwara, N., Takahashi, M., Miyamoto, Y., Koh, Y. H., Suzuki, K. & Taniguchi, N. (2005). Induction of thioredoxin reductase as an adaptive response to acrolein in human umbilical vein endothelial cells. *Biochem. Biophys. Res. Commun., 327*, 1058–1065.

Perluigi, M., Coccia, R. & Butterfield, D. A. (2012). 4-Hydroxy-2-nonenal, a reactive product of lipid peroxidation, and neurodegenerative diseases: a toxic combination illuminated by redox proteomics studies. *Antiox. Redox Sign., 17*, 1590–1609.

Petrović, S., Leskovac, A., Kotur-Stevuljević, J., Joksic, J., Guc-Scekic, M., Vujic, D. & Joksic, G. (2011). Gender-related differences in the oxidant state of cells in Fanconi anemia heterozygotes. *Biol. Chem., 392*, 625-632.

Pizzimenti, S., Ciamporcero, E., Daga, M., Pettazzoni, P., Arcaro, A., Cetrangolo, G., Minelli, R., Dianzani, C., Lepore, A., Gentile, F. & Barrera, G. (2013). Interaction of aldehydes derived from lipid peroxidation and membrane proteins. *Front. Physiol., 4*, 242.

Poot, M., Gross, O., Epe, B., Pflaum, M. & Hoehn, H. (1996). Cell cycle defect in connection with oxygen and iron sensitivity in Fanconi anemia lymphoblastoid cells. *Exp. Cell Res., 222*, 262-268.

Pryor, W. A. & Stanley, J. P. (1975). A suggested mechanism for the production of malonaldehyde during the autoxidation of polyunsaturated fatty acids. Nonenzymatic production of prostaglandin endoperoxides during autoxidation. *J. Org. Chem., 40*, 3615-3617.

Rahman, I., Van Schadewijk, A. A. M., Crowther, A. J. L., Hiemstra, P. S., Stolk, J., MacNee, W. & De Boer, W. I. (2002). 4-Hydroxy-2-nonenal, a specific lipid peroxidation product, is elevated in lungs of patients with chronic obstructive pulmonary disease. *Am. J. Respir. Crit.Care Med., 166*, 490–495.

Riahi, Y., Cohen, G., Shamni, O. & Sasson, S. (2010). Signaling and cytotoxic functions of 4-hydroxyalkenals. *Am. J. Physiol.: Endocrinol. Metabolism, 299*, E879–E886.

Ruppitsch, W., Meisslitzer, C., Hirsch-Kauffmann, M. & Schweiger, M. (1998). Overexpression of thioredoxin in Fanconi Anemia fibroblasts prevents the cytotoxic and DNA damaging effect of mitomycin C and diepoxybutane. *FEBS Lett., 422*, 99-102.

Sakata, K., Kashiwagi, K., Sharmin, S., Ueda, S. & Igarashi, K. (2003). Acrolein produced from polyamines as one of the uraemic toxins. *Biochem. Soc. Trans., 31*, 371–374.

Schaur, R. J. (2003). Basic aspects of the biochemical reactivity of 4-hydroxynonenal. *Mol. Aspects Med., 24*, 149–159.

Schneider, C., Tallman, K. A., Porter, N. A. & Brash, R. A. (2001). Two distinct pathways of formation of 4-hydroxynonenal.Mechanisms of nonenzymatic transformation of the 9- and 13-hydroperoxides of linoleic acid to 4-hydroxyalkenals. *J. Biol. Chem., 276*, 20831–20838.

Siems W. & Grune, T. (2003). Intracellular metabolism of 4-hydroxynonenal. *Mol. Asp. Med., 24*, 167–175.

Singh, M., Nam, D. T., Arseneault, M. & Ramassamy, C. (2010). Role of by-products of lipid oxidation in Alzheimer's disease brain: a focus on acrolein. *J. Alzheimers Dis., 21*, 741–756.

Slatter, D. A., Avery, N. C. & Bailey, A. J. (2004). Identification of a new cross-link and unique histidine adduct frombovine serum albumin incubated withmalondialdehyde. *J. Biol. Chem., 279*, 61–69.

Spickett, C. M. (2013). The lipid peroxidation product 4-hydroxy-2-nonenal: advances in chemistry and analysis. *Redox Biol., 1*, 145–152.

Spiteller, G., Kern, W. & Spiteller, P. (1999). Investigation of aldehydic lipid peroxidation products by gas chromatography-mass spectrometry. *J. Chrom. A, 843*, 29–98.

Stevens, J. F. & Maier, C. S. (2008). Acrolein: sources, metabolism, and biomolecular interactions relevant to human health and disease. *Mol. Nutr. Food Res., 52*, 7–25.

Stone, M. P., Cho, Y.-J., Huang, H., Kim, H.-Y., Kozekov, I. D., Kozekova, A., Wang, H., Minko, I. G., Lloyd, R. S., Harris, T. M. & Rizzo, C. J. (2008). Interstrand DNA crosslinks induced by α,β-unsaturated aldehydes derived from lipid peroxidation and environmental sources. *Acc. Chem. Res., 41*, 793–804.

Takeuchi, T. & Morimoto, K. (1993). Increased formation of 8-hydroxydeoxy-guanosine, an oxidative DNA damage, in lymphoblasts from Fanconi's anemia patients due to possible catalase deficiency. *Carcinogenesis, 14*, 1115-1120.

Tanuma, N., Miyata, R., Hayashi, R., Uchiyama, A. & Kurata, K. (2008). Oxidative stress as a biomarker of respiratory disturbance in patients with severe motor and intellectual disabilities. *Brain Dev., 30*, 402–409.

Thompson, C. A. & Burcham, P. C. (2008). Genome-wide transcriptional responses to acrolein. *Chem. Res. Toxicol., 21*, 2245–2256.

Thornalley, P. J. (1993). The glyoxalase system in health and disease. *Mol. Asp. Med., 14*, 287–371.

Thornalley, P. J. (1996). Pharmacology of methylglyoxal: formation, modification of proteins and nucleic acids, and enzymatic detoxification—a role in pathogenesis and antiproliferative chemotherapy. *Gen. Pharmacol., 27*, 565–573.

Thornalley, P. J., Langborg, A. & Minhas, H. S. (1999). Formation of glyoxal, methylglyoxal and 8-deoxyglucosone in the glycation of proteins by glucose. *Biochem. J., 344*, 109–116.

Traverso, N., Menini, S., Pesce Maineri, E., Patriarca, S., Odetti, P., Cottalasso, D., Marinari, U. M. & Pronzato, M. A. (2004). Malondialdehyde, a lipoperoxidation-derived aldehyde, can bring about secondary oxidative damage to proteins. *J. Gerontol. A: Biol. Sci. Med. Sci., 59*, 890–895.

Uchida, K., Kanematsu, M., Morimitsu, Y., Osawa, T., Noguchi, N. & Niki, E. (1998). Acrolein is a product of lipid peroxidation reaction. Formation of free acrolein and its conjugate with lysine residues in oxidized low density lipoproteins. *J. Biol. Chem., 273*, 16058–16066.

Uchida, K., Shiraishi, M., Naito, Y., Torii, Y., Nakamura, Y. & Osawa, T. (1999). Activation of stress signaling pathways by the end product of lipid peroxidation: 4-Hydroxy-2-nonenal is a potential inducer of intracellular peroxide production. *J. Biol. Chem., 274*, 2234–2242.

Uchida, K. (1999). Current status of acrolein as a lipid peroxidation product. *Trends Cardiovasc. Med., 9*, 109–113.

Urata, G. & Granick, S. (1963). Biosynthesis of α-Aminoketones and the Metabolism of Aminoacetone. *J. Biol. Chem., 238*, 811-820.

Usatyuk, P. V. & Natarajan, V. (2012). Hydroxyalkenals and oxidized phospholipids modulation of endothelial cytoskeleton, focal adhesion and adherens junction proteins in regulating endothelial barrier function. *Microvasc. Res., 83*, 45–55.

Van Der Toorn, M., Smit-de Vries, M. P., Slebos, D.-J., de Bruin, H. G., Abello, N., van Oosterhout, A. J. M., Bischoff, R. & Kauffman, H. F. (2007). Cigarette smoke irreversibly modifies glutathione in airway epithelial cells. *Am. J. Physiol., 293*, L1156–L1162.

VanderVeen, L. A., Hashim, M. F., Shyr, Y. & Marnett, L. J. (2003). Induction of frameshift and base pair substitution mutations by the major DNA adduct of the endogenous carcinogen malondialdehyde. *Proc. Nat. Acad. Sci. USA, 100*, 14247–14252.

Voehringer, M.-L., Becker, T. W., Krieger, G., Jacobi, H. & Witte, I. (1998). Synergistic DNA damaging effects of malondialdehyde/Cu(II) in PM2DNA and in human fibroblasts. *Toxicol. Lett., 94*, 159–166.

Voulgaridoua, G.-P., Anestopoulosa, I., Francob, R., Panayiotidisc, M. I. & Pappa, A. (2011). DNA damage induced by endogenous aldehydes: Current state of knowledge. *Mutat. Res., 711*, 13–27.

Wautier, J. L. & Schmidt, A. M. (2004). Protein glycation: a firm link to endothelial cell dysfunction. *Circ. Res., 95*, 233–8.

Weinhold, B. (2011). Acrolein and neuro disorders. *Environ. Health Perspect., 119*, a68.

Wells-Knecht, K. J., Zyzak, D. V., Litchfield, J. E., Thorpe, S. R. & Baynes, J. W. (1995). Mechanism of autoxidative glycosylation: identification of glyoxal and arabinose as intermediates in the autoxidative modification of proteins by glucose. *Biochemistry, 34*, 3702–3709.

Whitsett, J., Picklo Sr., M. J. & Vasquez-Vivar, J. (2007). 4-Hydroxy-2-nonenal increases superoxide anion radical in endothelial cells via stimulated GTP cyclohydrolase proteasomal degradation. *Arteriosc l. Thromb. Vasc. Biol., 27*, 2340–2347.

Willis, M. S., Klassen, L. W., Carlson, D. L., Brouse, C. F. & Thiele, G. M. (2004). Malondialdehyde-acetaldehyde haptenated protein binds macrophage scavenger receptor(s) and induces lysosomal damage. *Int. Immunopharmacol., 4*, 885–899.

Wolff, S. P., Jiang, Z. Y. & Hunt, J. W. (1991). Protein glycation and oxidative stress in diabetes mellitus and ageing. *Free Radic. Biol. Med., 10*, 339–52.

Wu, C. C., Hsieh, C. W., Lai, P. H., Lin, J. B., Liu, Y. C. & Wung, B. S. (2006). Upregulation of endothelial heme oxygenase-1 expression through the activation of the JNK pathway by sublethal concentrations of acrolein. *Toxicol. Appl. Pharmacol., 214*, 244–252.

Yin, H. & Porter, N. A. (2005). New insights regarding the autoxidation of polyunsaturated fatty acids. *Antiox. Redox Signal., 7*, 170–184.

Yun, M. R., Park, H. M., Seo, K. W., Lee, S. J., Im, D. S. & Kim, C. D. (2010). 5-Lipoxygenase plays an essential role in 4-HNE-enhanced ROS production in murine macrophages via activation of NADPH oxidase. *Free Radic. Res., 44*, 742–750.

Zarkovic, N., Cipak, A., Jaganjac, M., Borovic, S. & Zarkovic, K. (2013). Pathophysiological relevance of aldehydic protein modifications. *J. Proteomics, 92*, 239–247.

Zimniak, P. (2011). Relationship of electrophilic stress to aging. *Free Radic. Biol. Med., 51*,1087–1105.

In: Fanconi Anemia and Oxidative Stress
Editor: Giovanni Pagano
ISBN: 978-1-63482-297-8

Chapter 6

Role of Glutathione and Other Antioxidant Related Mechanisms in Fanconi Anemia

José Luis García Giménez, Carlos Romá-Mateo and Federico V. Pallardó*

Center for Biomedical Network Research on Rare Diseases. Biomedical Research Institute INCLIVA. Department of Physiology, Faculty of Medicine and Dentistry, University of Valencia, Spain

Abstract

Fanconi Anemia (FA; ORPHA84, CIE-10 D61.0) is a rare genetic disease characterized by congenital anomalies, hematological deficiencies and cancer predisposition. The family of FA proteins is involved in the responses against cellular endogenous damage and replicative stress, thus it has been proposed that oxidative damage, which has been found *in vivo* and *in vitro,* is a key component in the FA phenotype. Furthermore, redox-related functions have been found for the Fanconi proteins FANCC, FANCA, FANCG, FANCD2 and FANCJ. Although not completely known, the mechanisms by which FA proteins control the production of reactive oxygen species may relate to the altered mitochondrial function found in FA cells. Glutathione (GSH) is a very important antioxidant that has been found to be prone to oxidation in whole blood of FA patients. In addition, the functions of GSH go beyond its role as antioxidant, controlling the activity of transcription factors, epigenetic phenomena and cell proliferation, among others. Here, we describe the crosslink between GSH and other antioxidant mechanisms in the FA phenotype, and how the elucidation of this interaction can contribute to improve chemopreventive protocols for FA.

Keywords: Fanconi Anemia, glutathione, oxidative stress, antioxidant response elements, mitochondria, epigenetics, rare disease

---

*Corresponding Author: E-mail: Federico.V.Pallardó@uv.es.

Introduction

Fanconi Anemia (FA) (ORPHA84, CIE-10, D61.0) is a rare genetic disease with a prevalence of 1-9 per million and autosomal recessive or X-linked recessive inheritance, depending of the complementation group. Recent studies on carrier frequency gave an estimation of more than 1/200, with an expected prevalence at birth of at least 1/160.000 (Gulbis et al. 2010), although in certain populations the carrier frequency is higher due to founder mutation. FA is characterized by bone marrow failure leading to progressive pancytopenia, excess leukemia, and cancer risk, as well as a broad array of clinical complications and variable congenital malformations, present in 60%-75% of affected individuals, and predisposition to develop hematological or solid tumors (Alter 1993-2014). In this regard, by age 40 to 50 years, the estimated cumulative incidence of bone marrow failure is 90%; the incidence of hematologic malignancies (primarily acute myeloid leukemia) is 10-30%, and of non-hematologic malignancies and solid tumors, such as head and neck, skin, gastrointestinal tract, and genital tract, is of 25-30%.

Based on somatic cell fusion studies, 15 complementation FA groups have been identified; A, B, D1 (BRCA2), D2, E, F, G, I, J (BRIP1/BACH), L, M, N (PALB2), O (RAD51C), and P (SLX4). Recently, the group of Surrallés has identified a new complementation group (FANCQ) by whole-exome and Sanger sequencing in unclassified FA individuals, and discovered biallelic germline mutations in ERCC4 (XPF), a structure-specific nuclease-encoding gene described previously in Xeroderma pigmentosum (Bogliolo et al. 2013). All FA patients show, in common, hypersensitivity to the clastogenic effect of DNA cross-linking agents and hypersensitivity to oxidative damage.

Early reports already described the involvement of oxidative stress (OS) in FA: Nordenson in 1977 reported how spontaneous chromosomal instability in FA cells was mitigated by superoxide dismutase (SOD) (Nordenson 1977), and Joenje et al. reported the toxicity of higher pO_2 in FA cells (Joenje et al. 1981). Almost 40 years after these pioneering works, a body of literature has been accumulated, evidencing the role of OS in the physiopathology of FA.

In this regard, some publications have found that individual sensitivity to the genotoxic and cytotoxic effects of diepoxybutane (DEB), one of the main criteria to diagnose FA depend on oxygen levels and result in a dose-related reduced glutathione (GSH) depletion (Boogaard and Bond 1996).

Furthermore, GSH S-transferase (GST) and GSH peroxidase (GPx) activities are involved in the detoxification of DEB-induced DNA damage (Spanò et al. 1998; Erexson and Tindall 2000). GSH is synthesized in the cytosol; however, it is distributed to different intracellular compartments, where it performs different organelle-specific functions related to its different roles. These specific roles of GSH into each cellular organelle may have important implications in the pathophysiology of FA, corresponding to each specific complementation group.

Here we describe the role of GSH in different molecular pathways involved in the physiology of FA cells.

GSH Metabolism

GSH Synthesis, Homeostasis and Recycling

Glutathione is the primary low-molecular-weight thiol in all aerobic cells. The key functions of GSH, due to the presence of its cysteinyl residue, include: (1) working as an antioxidant to remove reactive oxygen species (ROS) and reactive nitrogen species (RNS); (2) acting as a reservoir of cysteine; (3) acting as a cofactor of antioxidant and anti-xenobiotic (Phase II) enzymes, including GSH *S*-transferases, GSH peroxidases, and glutaredoxins (Motohashi and Yamamoto 2004; Taguchi et al. 2011); (4) recycling of other antioxidants (such as ascorbic acid) (May et al. 1996); (5) protecting against inflammatory responses by modulating redox-regulated signal transduction and redox-sensitive transcription factors (NF-κB and AP1) (Rahman 2000; Markovic et al. 2010; García-Giménez et al. 2013a); (6) regulating cell proliferation (Markovic et al. 2007; Reddy et al. 2007); (7) protecting against apoptosis (Armstrong et al. 2002; Filomeni et al. 2003); and (8) regulating the structure of the nucleosome (García-Giménez et al. 2013b). Therefore, understanding the role of cellular levels of GSH or the regulation of several proteins in which GSH participates beyond its role as antioxidant, signaling molecule, or co-factor for the detoxification of xenobiotics, is essential to gain deep insight in the pathophysiology of GSH in health and disease.

The synthesis of GSH from its constituent amino acids is provided from the trans-sulfuration pathway and Krebs cycle. It involves the actions of two ATP-dependent enzymes, γ-glutamylcysteine ligase (GCL) and GSH synthetase (GS), both participating in the final steps of the GSH biosynthesis pathway.

Extracellular degradation of GSH is initiated by γ-glutamyl transpeptidase (GGT), which participates in the production of cysteine, in a rapid and irreversible step. GSH reductase (GR) reduces the oxidized form of glutathione, glutathione disulfide (GSSG), to the reduced form, GSH, using NADPH as a cofactor. Thus, GR is an important antioxidant enzyme and its deficiency produces a specific type of haemolytic anemia (ICD-10 D44.1; ORPHA90030), characterized by favism and cataracts and only described in three patients worldwide, born from consanguineous parents.

Other relevant antioxidant enzymes that use GSH as a cofactor are glutathione peroxidases (GPx). GPx are a widespread protein superfamily able to metabolize hydrogen peroxide and lipid hydroperoxides, using GSH as a co-substrate (Arthur, 2000; Toppo et al. 2008), and generating GSSG as a subproduct.

Glutaredoxins (Grxs) and thioredoxins (Trx) are thiol-disulfide oxidoreductases. In particular, Grxs are GSH-dependent enzymes involved in the maintenance of cellular redox homeostasis by protein deglutathionylation reactions and/or reducing disulfide bonds (Bushweller et al. 1992; Gravina and Mieyal 1993).

Finally, glutathione S-transferases (GST) are a family of enzymes that conjugate glutathione with electrophilic and hydrophobic compounds, and xenobiotics in general. Specifically, *GSTP1* gene expression is of particular interest in breast (Jhaveri and Morrow 1998) and prostate cancer (Nakayama et al. 2004). Interestingly, some investigations have described functional interaction of FANCC and FANCG proteins with NADPH cytochrome P450 reductase and in the redox regulation of GSTP1 (Kruyt et al. 1998; Futaki et al. 2002; Cumming et al. 2001).

Nrf2 Pathway

The nuclear factor erythroid-2-related factor 2 (Nrf2) is the principal transcription factor that regulates the antioxidant response element (ARE)-mediated gene transcription. On the other hand, the Kelch-associated protein 1 (Keap1) is a cysteine-rich protein that interacts with the factor Nrf2 in a redox-sensitive manner, blocking the migration of Nrf2 to the nucleus (Taguchi et al. 2011). More than 200 antioxidant and detoxification enzymes are regulated by the Keap1-Nrf2 pathway (Kwak et al. 2003); thus, the alteration of the nuclear interaction between Nrf2 and ARE gene clusters have critical effects on cellular antioxidant and xenobiotic responses. Interestingly, Keap1 contains abundant free cysteine residues (25 in mice and 27 in humans), which makes it a redox-sensing enzyme (Dinkova-Kostova et al. 2002): under OS conditions, it releases Nrf2, avoiding its degradation and allowing migration to the nucleus. In these conditions, Nrf2 in association with small proteins from the Maf and Jun families, forms an upstream transcriptional complex that binds to the ARE sequences of DNA, activating transcription of downstream genes such as NAD(P)H quinine oxidoreductase 1 (NQO-1), GSH S transferases (GSTs), UDP-glucosyl transferase 1-Ab, γ-glutamylcysteine ligase (GCL), heme oxygenase 1 (HO-1), thioredoxin reductase-1 (TrxR1), and thioredoxin (Trx) (Anuranjani and Bala 2014).

Fanconi Proteins, Nrf2 Pathway and Their Roles in GSH Metabolism

It is noteworthy that the BTB and CNC homology 1 basic leucine zipper transcription factor 1, also known as FANCJ/BACH1/BRIP1, competes with Nrf2, leading to a negative regulation of the ARE-mediated NADPH:quinone oxidoreductase 1 gene expression and induction in response to antioxidants (Dhakshinamoorthy et al. 2005). Furthermore, FANCJ/BACH1/BRIP1 regulates HO-1 (Kitamuro et al. 2003; Okada et al. 2010).

FANCJ/BACH1/BRIP1 is a transcriptional repressor ubiquitously expressed in tissues. Regarding OS response pathways, there exist three target genes, of which two showed marked expression changes after FANCJ/BACH1/BRIP1 knockdown (Warnatz et al. 2011). These target genes are the cystine/glutamate antiporter *SLC7A11* and the two subunits of GCL, GCLC and GCLM, in turn one of the key enzymes in the metabolism of the amino acids necessary for GSH synthesis. The decreased expression of the *SLC7A11* gene can lower the cellular uptake of oxidized cysteine, whereas reduced cysteine can still be internalized via the ubiquitous ASC transport system (Bussolati et al. 1992). Thus, the intracellular level of cysteine, which is important for glutathione synthesis, can be elevated. Accordingly, SLC7A11 is involved in glutathione-dependent neuroprotection from OS(Shih et al. 2006). Two more FANCJ/BACH1/BRIP1 targets involved in the OS response are the GCLC and GCLM subunits. Warnatz et al. reported that *GCLM* was strongly up-regulated after FANCJ/BACH1/BRIP1 knockdown, leading to an increased synthesis of glutathione (Warnatz et al. 2011).

Other FANC proteins have been related with the function proteins participating in the metabolism of GSH-metabolism related proteins. In particular, FANCC was found to interact directly with GSTP1 (Cumming et al. 2001). Cumming et al. described the role of FANCC preventing the formation of inactivating disulfide bonds within GSTP1, when apoptosis is activated. This represents an interesting role for FANCC, since this protein lacks homology with conventional disulfide reductases (Cumming et al. 2001). GSTP1 has been found in cytoplasm, mitochondria, and nucleus (Goto et al. 2009). This data are interesting and warrant further investigation, since FANCC has been described mainly in nucleus and cytoplasm, but not in mitochondria (Kupfer et al. 1997).

GSH and Mitochondria in Fanconi Anemia

It is noteworthy that the BTB and CNC homology 1 basic leucine zipper transcription factor 1, also known as FANCJ/BACH1/BRIP1, competes with Nrf2, leading to a negative regulation of the ARE-mediated NADPH:quinone oxidoreductase 1 gene expression and induction in response to antioxidants (Dhakshinamoorthy et al. 2005). Furthermore, FANCJ/BACH1/BRIP1 regulates HO-1 (Kitamuro et al. 2003; Okada et al. 2010).

FANCJ/BACH1/BRIP1 is a transcriptional repressor ubiquitously expressed in tissues. Regarding OS response pathways, there exist three target genes, of which two showed marked expression changes after FANCJ/BACH1/BRIP1 knockdown (Warnatz et al. 2011). These target genes are the cystine/glutamate antiporter *SLC7A11*and the two subunits of GCL, GCLC and GCLM, in turn one of the key enzymes in the metabolism of the amino acids necessary for GSH synthesis. The decreased expression of the *SLC7A11* gene can lower the cellular uptake of oxidized cysteine, whereas reduced cysteine can still be internalized via the ubiquitous ASC transport system (Bussolati et al. 1992). Thus, the intracellular level of cysteine, which is important for glutathione synthesis, can be elevated. Accordingly, SLC7A11 is involved in glutathione-dependent neuroprotection from OS (Shih et al. 2006). Two more FANCJ/BACH1/BRIP1 targets involved in the OS response are the GCLC and GCLM subunits. Warnatz et al. reported that *GCLM* was strongly up-regulated after FANCJ/BACH1/BRIP1 knockdown, leading to an increased synthesis of glutathione (Warnatz et al. 2011).

Other FANC proteins have been related with the function proteins participating in the metabolism of GSH-metabolism related proteins. In particular, FANCC was found to interact directly with GSTP1 by Cumming et al. (2001), who described the role of FANCC preventing the formation of inactivating disulfide bonds within GSTP1, when apoptosis was activated. This represents an interesting role for FANCC, since this protein lacks homology with conventional disulfide reductases (Cumming et al. 2001). GSTP1 has been found in cytoplasm, mitochondria, and nucleus (Goto et al. 2009). These data warrant further investigation, since FANCC has been described mainly in nucleus and cytoplasm, but not in mitochondria (Kupfer et al. 1997).

Nuclear GSH and DNA Repair in Fanconi Anemia

An interesting point regarding the interaction of Nrf2 with the ARE-genes is that Nrf2 gene expression is itself governed by the ARE (Kwak et al. 2002). This intriguing possibility exists since the deregulation that alters ARE-mediated gene transcription is produced, for example, during aging (Kwak et al. 2002), a physiological process characterized by depletion of GSH levels (Liu and Choi 2000; Rebrin et al. 2003; Sastre et al. 2000). The link between chromatin structure, accessibility of ARE-genes and FA is evident, since OS can alter the stability of the chromatin, affecting the recognition of ARE-genes by FA proteins. Furthermore, nuclear GSH is important for the integrity of the nuclear compartment, DNA protectionand also chromatin stability (García-Giménez et al. 2013a,b; Markovic et al. 2010).

It was reported that the Nrf2-ARE pathway overexpression alters the multidrug resistance associated protein (MRP) (Figure 1). MRPs are responsible for decreasing the drug susceptibility by reducing intracellular accumulation of drugs, in a process that depends on GSH (Maher et al. 2007). Overexpression of Nrf2 promotes the internalization of GSH within the nucleus by MRPs transporters, thereby avoiding the accumulation of chemotherapeutic drugs inside the nucleus (Ji et al. 2013). So, the overexpression of Nrf2 results in an increase in the concentration of cellular GSH, by means of regulating the expression of GCLC together with the internalization of GSH within the nucleus, to protect the DNA (Figure 1).

Truncated mutations of PALB2, also known as FANCN, have been identified in familial breast cancer and in the N complementation group of FA (Reid et al. 2007; Xia et al. 2007). Importantly, PALB2 is a major BRCA2 binding partner that controls its intracellular localization and stability, contributing to homologous recombination during repairing of double-strand breaks in DNA (Moynahan and Jasin 2010). PALB2/FANCN shares with Nrf2 a highly conserved ETGE-type Keap1 binding motif, so it can compete with Nrf2 for Keap1 binding within the nucleus (Ma et al. 2012). Therefore, PALB2/FANCN promotes Nrf2 accumulation and function in the nucleus to decrease the cellular ROS. PALB2 also regulates the rate of Nrf2 export from the nucleus following its induction (Ma et al. 2012) (Figure 1). It seems that a FANCN proper function must be critical to maintain the GSH levels inside the nucleus. Although there are no reported data in this regard, nuclear GSH measurement in these FANCN cells could clarify this hypothesis.

As described above, in the absence of cellular stress, FANCJ/BACH1/BRIP1 heterodimerizes with Maf proteins (MafF, MafG, and MafK) proteins that bind to the ARE, repressing the gene expression mediated by these elements. In presence of OS, FANCJ/BACH1/BRIP1 releases from the protein complex bound to ARE and allows the interaction of Nrf2. Thu, FANCJ/BACH1/BRIP1competes with Nrf2 for binding to the ARE, suppressing the expression of Nrf2 downstream genes (Dhakshinamoorthy et al. 2005) (Figure 1). Using directed mutagenesis experiments, a report marked the importance of cysteine residues C557 and C574 in regulating FANCJ/BACH1/BRIP1 protein function in human endothelial cells (Meng et al. 2010), thus suggesting that these cysteine residues may participate in the repression of ARE genes (Figure 1).

A possibility integrating those mechanisms mediated by PALB2/FANCN and FANCJ/BACH1/BRIP1 is that OS promotes the interaction of PALB2/FANCN with Nrf2, releasing Keap1 and thereby concentrating the complex PALB2/FANCN-Nrf2 into the nucleus. At the same time, OS may alter the oxidation state of C557 and C774 in FANCJ/BACH1/BRIP1, which in turn releases from the ARE genes. Therefore, this mechanism produces an optimal concentration of Nrf2 in order to bind to ARE, inducing GCL subunits expression and hence increasing the synthesis of GSH. In parallel, MRP and Nrf2 may facilitate the entrance of GSH within the nucleus, reducing the OS inside the nucleus and protecting DNA from damage (Figure 1).

On the other hand, GSH is not only needed to prevent DNA damage and chromatin integrity, but also for the correct function of DNA repairing enzymes. GSH is an essential molecule in DNA repair pathways. Oxidative lesions on DNA are principally repaired by the base excision repair (BER) pathway. Some of the participating proteins during BER are DNA glycosylases (such as 8-oxoguanine glycosylase, hOGG1), which remove the oxidized base creating an apurinic site, and AP endonucleases, which nick the damaged DNA on the 5' side of the apurinic/apyrimidinic (AP) sites.

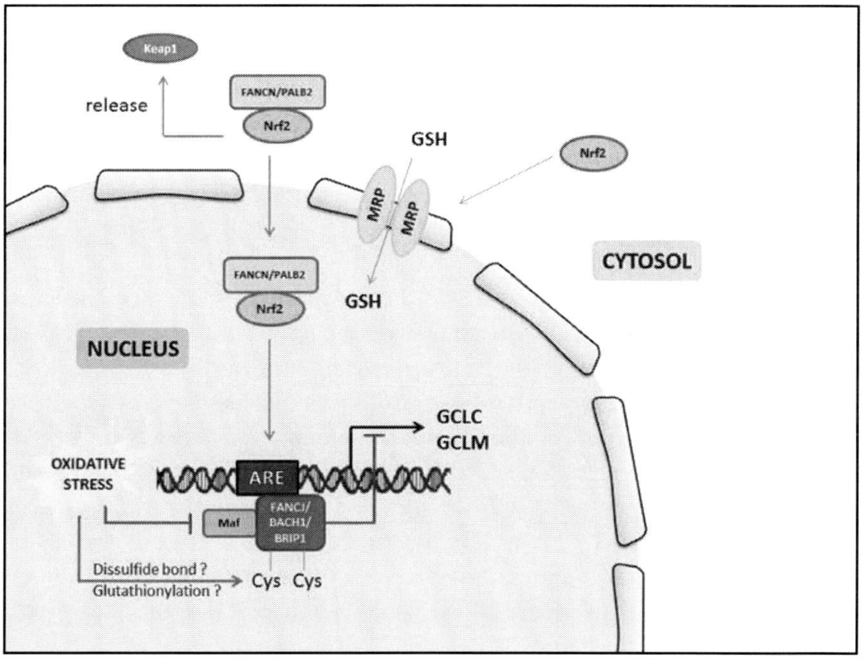

Figure 1. FANCN regulation of Nrf2-Keap1 mechanism.

In the cytosol, FANCN/PALB2 competes with Keap1 and facilitates Nrf2 accumulation into the nucleus, thus contributing to ARE genes transcription. Free Nrf2 also stimulates MRP transporters, facilitating GSH import. In the nucleus FANCJ/BACH1/BRIP1 heterodimerizes with Maf proteins that bind to the ARE, repressing the ARE-mediated gene expression. In presence of OS, FANCJ/BACH1/BRIP1 releases from the protein complex bound to ARE, allowing the interaction of Nrf2. FANCJ/BACH1/BRIP1 itself also competes with Nrf2 for binding to the ARE, suppressing the expression of Nrf2 downstream genes.

Evidence for the role of GSH on DNA repair appeared some years ago (Lai et al. 1989; Evans et al. 1984). Lai et al. showed how L-buthionine sulfoximine (L-BSO), a specific inhibitor of GCL activity, also produced partial inhibition of cisplatin-associated DNA damage. However, the addition of glutathione ethyl ester after BSO-treatment recovered the DNA repair capacity (Lai et al. 1989).

The importance of GSH in DNA repair is also patent in some studies in which some genes involved in DNA repair pathways have been related to GSH synthesis, in small-cell lung cancer (Sun et al. 2010). Moreover, nuclear erythroid-2-related factor-1 (Nrf1), which is a paralog of BACH2, with association with Nrf2 acts as a transcription factor responsible for the expression of multiple antioxidant systems (Venugopal and Jaiswal 1998). Furthermore, Nrf1 can suppress the expression of Xeroderma pigmentosum C (XPC), a protein involved in the nucleotide scission repair mechanism (NER). Importantly, Nrf1 enhances the expression of XPC by increasing glutathione levels, and the addition of GSH restores DNA repair capacity in cells in which Nrf1 was inhibited (Abd Elmageed et al. 2012).

Although GSH does not seem to be an effective radio-protector against the damage to DNA induced by X-rays, it regulates the DNA repair machinery (Pujari et al. 2009). Afterwards, DNA integrity is restored by DNA polymerase I, Pol I. However, during BER other type of enzymes also participate, such as Poly(ADP-ribose)polymerases (PARPs), which interact with different components of the BER pathway (De Vos et al. 2012) and also with the repairing proteins during non-homologous end joining (NHEJ) repairing pathways. PARP-1, which is activated in response to a DNA single strand break (SSB) lesion, synthesizes PAR polymers on acceptors proteinssuch as histone H1, at the DNA lesion site. Thereafter, poly(ADP-ribosyl)ation of H1 induces the recruitment of other factors involved in BER or NHEJ (Krishnakumar et al. 2008; Abd Elmageed et al. 2012). If single strand breaks are not repaired, they can be converted to double strand breaks (DSBs). In this situation phosphorylation of the histone variant H2AX takes place, forming gamma-H2AX. The phosphorylated protein, gamma-H2AX, is the first step in recruiting and localizing DNA repair proteins. We have found that Poly(ADP)-ribosylation of histones not only occurred at all stages of cell cycle (Tanuma et al. 1978) but also it was highest in the active phase of cell proliferation, just when GSH concentration increases within the nucleus (García-Giménez et al. 2012). Therefore, the concentration of GSH could be very important for the activity of PARPs and mediate their recruitment into the nucleus, as we have previously described in plant cells (Diaz Vivancos et al. 2010; García-Giménez et al. 2013a; Pellny et al. 2009; Vivancos et al. 2010).

In fact, the glutathione pool increased in parallel with poly (ADP-ribose) polymerase (PARP) gene expression and activities, when cells are in active phase of cell proliferation (Pellny et al. 2009). The homology of these mechanisms in both plant and rodent cells, make us to think about this possibility in human cells (Pellny et al. 2009; García-Giménez et al. 2013b; Markovic et al. 2010).

In addition, it has been reported that PARP is depleted in fibroblasts obtained from Fanconi Anemia patients cultured at confluence (Ramirez et al. 2003), and PARP activity decreases when cells reach confluence (García-Giménez et al. 2012).

On the other hand, Lyakovich and Surrallés have evaluated PARP levels in fibroblasts obtained from the most severe complemented group of FA (FANCD2), comparing it with other FA complementation groups (Lyakhovich et al. 2011). They observed that FANCD2 safeguards genomic stability by inhibiting the parsylation activity of Tankyrase-1 (PARP)-mediated TRF1 (TNKS1) and by protecting telomeric DNA by TRF1 (a telomeric protein). This effect is produced because TNKS1 elimination blocks the PARsylation of TRF1, which in turn is liberated from the telomere, contributing to telomere shortening (Figure 2). In that way, TNKS1 may be in its inactivated form in FANCD2 corrected cells (in which FANCD2 is functional) but not in FANCD2 deficient cells. (Lyakhovich et al. 2011). In fact, it was reported that the knockdown of FANCD2 rapidly causes telomere dysfunction in cells that rely on alternative mechanisms of telomere lengthening (Spardy et al. 2008). Therefore, as Lyakovich suggests, TNKS1 may modulate telomere maintenance by protecting telomeric DNA. Besides, FANCD2 modulates TNKS1 activity, finally leading to the PARsylation of TRF1 and thus affecting the stability, but not the length, of telomeres. Finally, our group also described that telomerase activity decreases when GSH is depleted, and increases when GSH is replenished (Borrás et al. 2004). These results link all these mechanisms, in which PARP activity increases when GSH levels are high during S-phase, and FANCD2 activity is necessary for TNKS1 inhibition and subsequent TRF1 release from telomeres, which facilitates the telomere elongation during S-phase.

GSH Replenishment As Chemopreventive for FA

It is noteworthy that GSH levels are downregulated in FA patients (Pagano et al. 2004). For this reason, molecules able to replenish GSH levels and equilibrate the mitochondrial homeostasis in FA may be expected to act as chemopreventive agents to improve FA clinical evolution. In addition, increased oxidized glutathione ratio (GSSG/GSH) in plasma from FA patients has been found (Pagano et al. 2004).

Some studies have tested the use of different antioxidants for decreasing chromosome aberrations in FA cells. In that way, Dallapiccola et al. obtained a partial correction of chromosome damage in FA lymphocytes treated with antioxidants such as sodium ascorbate, L-cysteine, 2-mercaptoethanol, α-mercaptopropionylglycine, and GSH (Dallapiccola et al. 1985). Some of these molecules normalize GSH balance. In this regard, Ponte et al. (2012) proposed some strategies to replenish GSH cellular levels, for example using α-lipoic acid.

FANCD2 safeguards genomic stability by inhibiting the parsylation activity of Tankyrase-1 (PARP)-mediated TRF1 (TNKS1) and by protecting telomeric DNA by TRF1 (a telomeric protein). When nuclear GSH levels rise, non-mutated FANCD2 blocks TNKS1, maintaining TRF1 bound to the telomere. At low GSH concentrations and in the presence of a dysfunctional FANCD2, TNKS1 produces parsylation of TRF1, which is liberated from the telomere contributing to telomere instability.

Alpha-lipoic acid (ALA) is a natural and endogenous compound that is an essential cofactor for many mitochondrial enzymes. ALA has a cyclic disulfide moiety in redox equilibrium with its reduced dithiol form (dihydrolipoic acid), which in turn contributes to GSH replenishment, by reducing vitamin C, vitamin E, and coenzyme Q10 (Han et al. 1995; 1997; Tarnopolsky, 2008). Importantly, it has been shown that, in aged rats, ALA acts as a potent chemopreventive agent that increases cellular GSH and Phase II response genes due to its ability to induce Nrf2 binding to the ARE, and consequently, to activate the expression of GCLC and GCLM (Suh et al. 2004). NAC, the acetylated variant of L-cysteine, is a source of sulfhydryl groups, and it is converted into metabolites required for the synthesis of GSH (Kelly 1998). In fact, Ponte et al. (2012) tested the effect of 20 μM ALA and/or 500 μM NAC in DEB-treated lymphocyte cultures from FA patients. This study demonstrated that the concomitant exposure to ALA and NAC drastically improved the genetic stability in lymphocytes from FA patients *in vitro*.

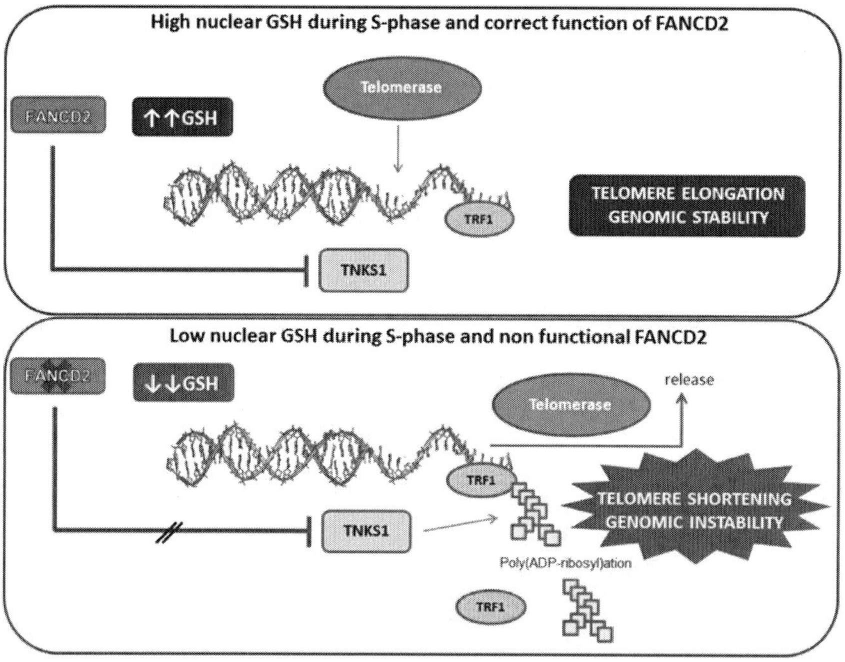

Figure 2. FANCD2 and TNKS1 interaction regulates telomere elongation and genomic stability.

The results reviewed here show that GSH, a key antioxidant in eukaryotic cells, plays an important function in the physiopathology of FA, a devastating disease in which a chronic pro-oxidative process takes place. Implementing molecular mechanisms addressed to decrease generation of free radicals by mitochondria or counteracting their deleterious effects could open new therapeutic possibilities in the fight against Fanconi Anemia.

References

Abd Elmageed, Z. Y., Naura, A. S., Errami, Y. & Zerfaoui, M. (2012). The poly(ADP-ribose) polymerases (PARPs): new roles in intracellular transport. *Cell Signal., 24*, 1-8.

Alter, B. P. & Kupfer, G. (1993-2014). Fanconi Anemia (2014 ed.): University of Washington, Seattle.

Anuranjani & Bala, M. (2014). Concerted action of Nrf2-ARE pathway, MRN complex, HMGB1 and inflammatory cytokines - implication in modification of radiation damage. *Redox Biol*, 2, 832-846.

Armstrong, J. S., Steinauer, K. K., Hornung, B., Irish, J. M., Lecane, P., Birrell, G. W. & Knox, S. J. (2002). Role of glutathione depletion and reactive oxygen species generation in apoptotic signaling in a human B lymphoma cell line. *Cell Death Differ., 9*, 252-263.

Arthur, J. R. (2000). The glutathione peroxidases. *Cell Mol. Life Sci., 57*, 1825-1835.

Bogliolo, M., Schuster, B., Stoepker, C., Derkunt, B., Su, Y., Raams, A. & Surralles, J. (2013). Mutations in ERCC4, encoding the DNA-repair endonuclease XPF, cause Fanconi anemia. *Am. J. Hum. Genet., 92*, 800-806.

Boogaard, P. J. & Bond, J. A. (1996). The role of hydrolysis in the detoxification of 1,2:3,4-diepoxybutane by human, rat, and mouse liver and lung in vitro. Toxicol. *Appl. Pharmacol., 141*, 617-627.

Borrás, C., Esteve, J. M., Viña, J. R., Sastre, J., Viña, J. & Pallardó, F. V. (2004). Glutathione regulates telomerase activity in 3T3 fibroblasts. *J. Biol. Chem., 279*, 34332-34335.

Bushweller, J. H., Aslund, F., Wuthrich, K. & Holmgren, A. (1992). Structural and functional characterization of the mutant Escherichia coli glutaredoxin and its mixed disulfide with glutathione. *Biochemistry, 31*, 9288-9293.

Bussolati, O., Laris, P. C., Rotoli, B. M., Dall'Asta, V. & Gazzola, G. C. (1992). Transport system ASC for neutral amino acids. An electroneutral sodium/amino acid cotransport sensitive to the membrane potential. *J. Biol. Chem., 267*, 8330-8335.

Cumming, R. C., Lightfoot, J., Beard, K., Youssoufian, H., O'Brien, P. J. & Buchwald, M. (2001). Fanconi anemia group C protein prevents apoptosis in hematopoietic cells through redox regulation of GSTP1. *Nat. Med,. 7*, 814-820.

Dallapiccola, B., Porfirio, B., Mokini, V., Alimena, G., Isacchi, G. & Gandini, E. (1985). Effect of oxidants and antioxidants on chromosomal breakage in Fanconi anemia lymphocytes. *Hum. Genet., 69*, 62-65.

De Vos, M., Schreiber, V. & Dantzer, F. (2012). The diverse roles and clinical relevance of PARPs in DNA damage repair: current state of the art. *Biochem. Pharmacol., 84*, 137-146.

Dhakshinamoorthy, S., Jain, A. K., Bloom, D. A. & Jaiswal, A. K. (2005). Bach1 competes with Nrf2 leading to negative regulation of the antioxidant response element (ARE)-mediated NAD(P)H:quinone oxidoreductase 1 gene expression and induction in response to antioxidants. *J. Biol. Chem., 280*, 16891-16900.

Diaz Vivancos, P., Wolff, T., Markovic, J., Pallardó, F. V. & Foyer, C. H. (2010). A nuclear glutathione cycle within the cell cycle. *Biochem. J., 431*, 169-178.

Dinkova-Kostova, A. T., Holtzclaw, W. D., Cole, R. N., Itoh, K., Wakabayashi, N., Katoh, Y. & Talalay, P. (2002). Direct evidence that sulfhydryl groups of Keap1 are the sensors regulating induction of phase 2 enzymes that protect against carcinogens and oxidants. *Proc. Natl. Acad. Sci. U S A, 99*, 11908-11913.

Erexson, G. L. & Tindall, K. R. (2000). Reduction of diepoxybutane-induced sister chromatid exchanges by glutathione peroxidase and erythrocytes in transgenic Big Blue mouse and rat fibroblasts. *Mutat. Res., 447*, 267-274.

Evans, J. W., Taylor, Y. C. & Brown, J. M. (1984). The role of glutathione and DNA strand break repair in determining the shoulder of the radiation survival curve. *Br. J. Cancer, Suppl. 6*, 49-53.

Filomeni, G., Rotilio, G. & Ciriolo, M. R. (2003). Glutathione disulfide induces apoptosis in U937 cells by a redox-mediated p38 MAP kinase pathway. *FASEB J., 17*, 64-66.

Futaki, M., Igarashi, T., Watanabe, S., Kajigaya, S., Tatsuguchi, A., Wang, J. & Liu, J. M. (2002). The FANCG Fanconi anemia protein interacts with CYP2E1: possible role in protection against oxidative DNA damage. *Carcinogenesis, 23*, 67-72.

García-Giménez, J. L., Ledesma, A. M., Esmoris, I., Romá-Mateo, C., Sanz, P., Viña, J. & Pallardó, F. V. (2012). Histone carbonylation occurs in proliferating cells. *Free Radic. Biol. Med., 52*, 1453-1464.

García-Giménez, J. L., Markovic, J., Dasi, F., Queval, G., Schnaubelt, D., Foyer, C. H. & Pallardó, F. V. (2013a). Nuclear glutathione. *Biochim. Biophys. Acta, 1830*, 3304-3316.

García-Giménez, J. L., Olaso, G., Hake, S. B., Bonisch, C., Wiedemann, S. M., Markovic, J. & Pallardó, F. V. (2013b). Histone h3 glutathionylation in proliferating mammalian cells destabilizes nucleosomal structure. *Antioxid. Redox Signal,. 19*, 1305-1320.

Goto, S., Kawakatsu, M., Izumi, S., Urata, Y., Kageyama, K., Ihara, Y. & Kondo, T. (2009). Glutathione S-transferase pi localizes in mitochondria and protects against oxidative stress. *Free Radic. Biol. Med., 46*, 1392-1403.

Gravina, S.A., Mieyal, J.J. (1993). Thioltransferase is a specific glutathionyl mixed disulfide oxidoreductase. *Biochemistry, 32*, 3368-3376.

Gulbis, B., Eleftheriou, A., Angastiniotis, M., Ball, S., Surralles, J., Castella, M. & Corrons, J. L. (2010). Epidemiology of rare anaemias in Europe. *Adv. Exp. Med. Biol., 686*, 375-396.

Han, D., Handelman, G., Marcocci, L., Sen, C. K., Roy, S., Kobuchi, H. & Packer, L. (1997). Lipoic acid increases de novo synthesis of cellular glutathione by improving cystine utilization. *Biofactors, 6*, 321-338.

Han, D., Tritschler, H. J. & Packer, L. (1995). Alpha-lipoic acid increases intracellular glutathione in a human T-lymphocyte Jurkat cell line. *Biochem. Biophys. Res. Commun., 207*, 258-264.

Hollins, D. L., Suliman, H. B., Piantadosi, C. A. & Carraway, M. S. (2006). Glutathione regulates susceptibility to oxidant-induced mitochondrial DNA damage in human lymphocytes. *Free Radic. Biol. Med. 40*, 1220-1226.

Jhaveri, M. S. & Morrow, C. S. (1998). Contribution of proximal promoter elements to the regulation of basal and differential glutathione S-transferase P1 gene expression in human breast cancer cells. *Biochim. Biophys. Acta, 1396*, 179-190.

Ji, L., Li, H., Gao, P., Shang, G., Zhang, D. D., Zhang, N. & Jiang, T. (2013). Nrf2 pathway regulates multidrug-resistance-associated protein 1 in small cell lung cancer. *PLoS One, 8*, e63404.

Joenje, H., Arwert, F., Eriksson, A. W., de Koning, H. & Oostra, A. B. (1981). Oxygen-dependence of chromosomal aberrations in Fanconi's anaemia. *Nature, 290*, 142-143.

Kelly, G. S. (1998). Clinical applications of N-acetylcysteine. *Altern. Med. Rev., 3*, 114-127.

Kitamuro, T., Takahashi, K., Ogawa, K., Udono-Fujimori, R., Takeda, K., Furuyama, K. & Shibahara, S. (2003). Bach1 functions as a hypoxia-inducible repressor for the heme oxygenase-1 gene in human cells. *J. Biol. Chem., 278*, 9125-9133.

Krishnakumar, R., Gamble, M. J., Frizzell, K. M., Berrocal, J. G., Kininis, M. & Kraus, W. L. (2008). Reciprocal binding of PARP-1 and histone H1 at promoters specifies transcriptional outcomes. *Science, 319*, 819-821.

Kruyt, F. A., Hoshino, T., Liu, J. M., Joseph, P., Jaiswal, A. K. & Youssoufian, H. (1998). Abnormal microsomal detoxification implicated in Fanconi anemia group C by interaction of the FAC protein with NADPH cytochrome P450 reductase. *Blood, 92*, 3050-3056.

Kumari, U., Ya Jun, W., Huat Bay, B. & Lyakhovich, A. (2014). Evidence of mitochondrial dysfunction and impaired ROS detoxifying machinery in Fanconi anemia cells. *Oncogene, 33*, 165-172.

Kupfer, G. M., Naf, D., Suliman, A., Pulsipher, M. & D'Andrea, A. D. (1997). The Fanconi anaemia proteins, FAA and FAC, interact to form a nuclear complex. *Nat. Genet., 17*, 487-490.

Kwak, M. K., Itoh, K., Yamamoto, M. & Kensler, T. W. (2002). Enhanced expression of the transcription factor Nrf2 by cancer chemopreventive agents: role of antioxidant response element-like sequences in the nrf2 promoter. *Mol. Cell. Biol., 22*, 2883-2892.

Kwak, M. K., Wakabayashi, N., Itoh, K., Motohashi, H., Yamamoto, M. & Kensler, T. W. (2003). Modulation of gene expression by cancer chemopreventive dithiolethiones through the Keap1-Nrf2 pathway. Identification of novel gene clusters for cell survival. *J. Biol. Chem., 278*, 8135-8145.

Lai, G. M., Ozols, R. F., Young, R. C. & Hamilton, T. C. (1989). Effect of glutathione on DNA repair in cisplatin-resistant human ovarian cancer cell lines. *J. Natl. Cancer Inst., 81*, 535-539.

Lash, L. H., Putt, D. A. & Matherly, L. H. (2002). Protection of NRK-52E cells, a rat renal proximal tubular cell line, from chemical-induced apoptosis by overexpression of a mitochondrial glutathione transporter. *J. Pharmacol. Exp. Ther., 303*, 476-486.

Liu, R. & Choi, J. (2000). Age-associated decline in gamma-glutamylcysteine synthetase gene expression in rats. *Free Radic. Biol. Med., 28*, 566-574.

Lyakhovich, A., Ramirez, M. J., Castellanos, A., Castella, M., Simons, A. M., Parvin, J. D. & Surralles, J. (2011). Fanconi anemia protein FANCD2 inhibits TRF1 polyADP-ribosylation through tankyrase1-dependent manner. *Genome Integr., 2*, 4.

Ma, J., Cai, H., Wu, T., Sobhian, B., Huo, Y., Alcivar, A. & Xia, B. (2012). PALB2 interacts with KEAP1 to promote NRF2 nuclear accumulation and function. *Mol. Cell. Biol., 32*, 1506-1517.

Maher, J. M., Dieter, M. Z., Aleksunes, L. M., Slitt, A. L., Guo, G., Tanaka, Y. & Klaassen, C. D. (2007). Oxidative and electrophilic stress induces multidrug resistance-associated protein transporters via the nuclear factor-E2-related factor-2 transcriptional pathway. *Hepatology, 46*, 1597-1610.

Mari, M., Morales, A., Colell, A., Garcia-Ruiz, C. & Fernandez-Checa, J. C. (2009). Mitochondrial glutathione, a key survival antioxidant. *Antioxid. Redox Signal., 11*, 2685-2700.

Markovic, J., Borrás, C., Ortega, A., Sastre, J., Viña, J. & Pallardó, F. V. (2007). Glutathione is recruited into the nucleus in early phases of cell proliferation. *J. Biol. Chem., 282*, 20416-20424.

Markovic, J., García-Giménez, J. L., Gimeno, A., Viña, J. & Pallardó, F. V. (2010). Role of glutathione in cell nucleus. *Free. Radic. Res., 44*, 721-733.

May, J. M., Qu, Z. C., Whitesell, R. R. & Cobb, C. E. (1996). Ascorbate recycling in human erythrocytes: role of GSH in reducing dehydroascorbate. *Free Radic. Biol. Med., 20*, 543-551.

Meng, D., Wang, X., Chang, Q., Hitron, A., Zhang, Z., Xu, M., Shi, X. (2010). Arsenic promotes angiogenesis in vitro via a heme oxygenase-1-dependent mechanism. *Toxicol. Appl. Pharmacol., 244*, 291-299.

Motohashi, H. & Yamamoto, M. (2004). Nrf2-Keap1 defines a physiologically important stress response mechanism. *Trends Mol. Med., 10*, 549-557.

Moynahan, M. E. & Jasin, M. (2010). Mitotic homologous recombination maintains genomic stability and suppresses tumorigenesis. *Nat. Rev. Mol. Cell Biol., 11*, 196-207.

Mukhopadhyay, S. S., Leung, K. S., Hicks, M. J., Hastings, P. J., Youssoufian, H. & Plon, S. E. (2006). Defective mitochondrial peroxiredoxin-3 results in sensitivity to oxidative stress in Fanconi anemia. *J. Cell Biol., 175*, 225-235.

Nakayama, M., Gonzalgo, M. L., Yegnasubramanian, S., Lin, X., De Marzo, A. M. & Nelson, W. G. (2004). GSTP1 CpG island hypermethylation as a molecular biomarker for prostate cancer. *J. Cell Biochem., 91*, 540-552.

Nordenson, I. (1977). Effect of superoxide dismutase and catalase on spontaneously occurring chromosome breaks in patients with Fanconi's anemia. *Hereditas, 86*, 147-150.

Okada, S., Muto, A., Ogawa, E., Nakanome, A., Katoh, Y., Ikawa, S. & Okuyama, R. (2010). Bach1-dependent and -independent regulation of heme oxygenase-1 in keratinocytes. *J. Biol. Chem., 285*, 23581-23589.

Pagano, G., Degan, P., d'Ischia, M., Kelly, F. J., Pallardó, F. V., Zatterale, A. & Warnau, M. (2004). Gender- and age-related distinctions for the in vivo prooxidant state in Fanconi anaemia patients. *Carcinogenesis, 25*, 1899-1909.

Pagano, G., Talamanca, A. A., Castello, G., d'Ischia, M., Pallardó, F. V., Petrović, S. & Zatterale, A. (2013). Bone marrow cell transcripts from Fanconi anaemia patients reveal in vivo alterations in mitochondrial, redox and DNA repair pathways. *Eur. J. Haematol., 91*, 141-151.

Palmieri, F. (1994). Mitochondrial carrier proteins. *FEBS Lett., 346*, 48-54.

Pellny, T. K., Locato, V., Vivancos, P. D., Markovic, J., De Gara, L., Pallardó, F. V. & Foyer, C. H. (2009). Pyridine nucleotide cycling and control of intracellular redox state in relation to poly (ADP-ribose) polymerase activity and nuclear localization of glutathione during exponential growth of Arabidopsis cells in culture. *Mol. Plant, 2*, 442-456.

Ponte, F., Sousa, R., Fernandes, A. P., Goncalves, C., Barbot, J., Carvalho, F. & Porto, B. (2012). Improvement of genetic stability in lymphocytes from Fanconi anemia patients through the combined effect of alpha-lipoic acid and N-acetylcysteine. *Orphanet J. Rare Dis., 7*, 28.

Pujari, G., Berni, A., Palitti, F. & Chatterjee, A. (2009). Influence of glutathione levels on radiation-induced chromosomal DNA damage and repair in human peripheral lymphocytes. *Mutat. Res., 675*, 23-28.

Rahman, I. (2000). Regulation of nuclear factor-kappa B, activator protein-1, and glutathione levels by tumor necrosis factor-alpha and dexamethasone in alveolar epithelial cells. *Biochem. Pharmacol., 60*, 1041-1049.

Ramirez, M. H., Adelfalk, C., Kontou, M., Hirsch-Kauffmann, M. & Schweiger, M. (2003). The cellular control enzyme polyADP ribosyl transferase is eliminated in cultured Fanconi anemia fibroblasts at confluency. *Biol. Chem., 384*, 169-174.

Ravera, S., Vaccaro, D., Cuccarolo, P., Columbaro, M., Capanni, C., Bartolucci, M. & Degan, P. (2013). Mitochondrial respiratory chain Complex I defects in Fanconi anemia complementation group A. *Biochimie, 95*, 1828-1837.

Rebrin, I., Kamzalov, S., Sohal, R. S. (2003). Effects of age and caloric restriction on glutathione redox state in mice. *Free Radic. Biol. Med., 35*, 626-635.

Reddy, N. M., Kleeberger, S. R., Cho, H. Y., Yamamoto, M., Kensler, T. W., Biswal, S. & Reddy, S. P. (2007). Deficiency in Nrf2-GSH signaling impairs type II cell growth and enhances sensitivity to oxidants. *Am. J. Respir. Cell. Mol. Biol., 37*, 3-8.

Reid, S., Schindler, D., Hanenberg, H., Barker, K., Hanks, S., Kalb, R. & Rahman, N. (2007). Biallelic mutations in PALB2 cause Fanconi anemia subtype FA-N and predispose to childhood cancer. *Nat. Genet., 39*, 162-164.

Ruppitsch, W., Meisslitzer, C., Hirsch-Kauffmann, M. & Schweiger, M. (1998). Overexpression of thioredoxin in Fanconi anemia fibroblasts prevents the cytotoxic and DNA damaging effect of mitomycin C and diepoxybutane. *FEBS Lett., 422*, 99-102.

Sastre, J., Pallardó, F. V., Garcia de la Asuncion, J. & Viña, J. (2000). Mitochondria, oxidative stress and aging. *Free Radic. Res., 32*, 189-198.

Shih, A. Y., Erb, H., Sun, X., Toda, S., Kalivas, P. W. & Murphy, T. H. (2006). Cystine/glutamate exchange modulates glutathione supply for neuroprotection from oxidative stress and cell proliferation. *J. Neurosci., 26*, 10514-10523.

Spanò, M., Cordelli, E., Leter, G. & Pacchierotti, F. (1998). Diepoxybutane cytotoxicity on mouse germ cells is enhanced by in vivo glutathione depletion: a flow cytometric approach. *Mutat. Res., 397*, 37-43.

Spardy, N., Duensing, A., Hoskins, E. E., Wells, S. I. & Duensing, S. (2008). HPV-16 E7 reveals a link between DNA replication stress, Fanconi anemia D2 protein, and alternative lengthening of telomere-associated promyelocytic leukemia bodies. *Cancer Res., 68*, 9954-9963.

Suh, J. H., Shenvi, S. V., Dixon, B. M., Liu, H., Jaiswal, A. K., Liu, R. M. & Hagen, T. M. (2004). Decline in transcriptional activity of Nrf2 causes age-related loss of glutathione synthesis, which is reversible with lipoic acid. *Proc. Natl. Acad. Sci. U S A, 101*, 3381-3386.

Sun, Z., Chen, J., Aakre, J., Marks, R. S., Garces, Y. Y., Jiang, R. & Yang, P. (2010). Genetic variation in glutathione metabolism and DNA repair genes predicts survival of small-cell lung cancer patients. *Ann. Oncol., 21*, 2011-2016.

Taguchi, K., Motohashi, H., Yamamoto, M. (2011). Molecular mechanisms of the Keap1-Nrf2 pathway in stress response and cancer evolution. *Genes Cells, 16*, 123-140.

Tanuma, S. I., Enomoto, T. & Yamada, M. A. (1978). Changes in the level of poly ADP-ribosylation during a cell cycle. *Exp. Cell Res., 117*, 421-430.

Tarnopolsky, M. A. (2008). The mitochondrial cocktail: rationale for combined nutraceutical therapy in mitochondrial cytopathies. *Adv. Drug Deliv. Rev., 60*, 1561-1567.

Toppo, S., Vanin, S., Bosello, V. & Tosatto, S. C. (2008). Evolutionary and structural insights into the multifaceted glutathione peroxidase (Gpx) superfamily. *Antioxid. Redox Signal., 10*, 1501-1514.

Venugopal, R. & Jaiswal, A. K. (1998). Nrf2 and Nrf1 in association with Jun proteins regulate antioxidant response element-mediated expression and coordinated induction of genes encoding detoxifying enzymes. *Oncogene, 17*, 3145-3156.

Vivancos, P. D., Dong, Y., Ziegler, K., Markovic, J., Pallardó, F. V., Pellny, T. K. & Foyer, C. H. (2010). Recruitment of glutathione into the nucleus during cell proliferation adjusts whole-cell redox homeostasis in Arabidopsis thaliana and lowers the oxidative defence shield. *Plant J., 64*, 825-838.

Warnatz, H. J., Schmidt, D., Manke, T., Piccini, I., Sultan, M., Borodina, T. & Yaspo, M. L. (2011). The BTB and CNC homology 1 (BACH1) target genes are involved in the oxidative stress response and in control of the cell cycle. *J. Biol. Chem., 286*, 23521-23532.

Xia, B., Dorsman, J.C., Ameziane, N., de Vries, Y., Rooimans, M.A., Sheng, Q., de Winter, J.P. (2007). Fanconi anemia is associated with a defect in the BRCA2 partner PALB2. *Nat. Genet., 39*, 159-161.

Zhang, H., Go, Y. M., Jones, D.P. (2007). Mitochondrial thioredoxin-2/peroxiredoxin-3 system functions in parallel with mitochondrial GSH system in protection against oxidative stress. *Arch. Biochem. Biophys., 465*, 119-126.

In: Fanconi Anemia and Oxidative Stress
Editor: Giovanni Pagano
ISBN: 978-1-63482-297-8

Chapter 7

Proinflammatory Cytokines As Mediators of Oxidative Stress in Fanconi Anemia

Annarita Aiello Talamanca* and Giuseppe Castello
CROM, National Cancer Institute "Fondazione Pascale" - IRCCS
Naples, Italy

Abstract

The most important immune defect in Fanconi Anemia (FA) is pancytopenia. The immune status in FA patients is characterized by altered IgG, IgM and IgA serum levels, a relative increase of cytotoxic T cells, naïve T cells, and regulatory T cells, higher levels of proinflammatory cytokines, in particular TNF-α, TGF-β, (IL)-6, INF-γ and IL-1β (IL-1β only in FA group A) and a lower number of Natural Killer (NK) cells compared to healthy donors.

Excess TNF-α levels may affect Hematopoietic Stem Cell (HSC) function directly by impairing HSC survival, homing and proliferation, or indirectly by changing the bone marrow microenvironment that is critical for HSC homeostasis and function; in this way it contributes to disease progression in FA. TNF-α-induced senescence seems to be strictly correlated with the accumulation of Reactive Oxygen Species (ROS) and oxidative DNA damage. Moreover, loss of FA proteins could be responsible for an upregulation of the basal NF-κB transcriptional activity, and therefore of some of the genes regulated by this pathway, for example, pro-inflammatory cytokines.

The redox-related pro-inflammatory state in FA cells has three important consequences: 1. the inhibition of HSC expansion; 2. increase of chromosomal instability predisposing to cancer; 3. the apoptosis activation in FA cells.

In this chapter we discuss the links between two of the most important hallmarks of FA, namely inflammation and oxidative stress (OS), and possible implications for the use of antioxidants to stop the initiation and delay disease progression.

*E-mail: a.aiello@istitutotumori.na.it; beppe.castello@gmail.com.

Keywords: inflammation, Tumor Necrosis Factor-α, Interferon-γ, Reactive Oxygen Species, Hematopoietic Stem Cells

Introduction

Fanconi Anemia patients with advanced marrow failure (before HSC transplantation) have increased serum levels of the myelosuppressive (TGF-β) and proinflammatory (interleukin (IL)-6) cytokines when compared to controls (Korthof et al. 2013).

Excess levels of the proinflammatory cytokine interleukin-1β (IL-1β) were detected in the serum from FA patients of complementation group FancA (Ibañez et al. 2009) while TNF-α and IFN-γ are overexpressed in bone marrow (Dufour et al. 2003) of FA patients. High level of TGF-β and IL-6 are in accordance with some characteristics of FA such as the proinflammatory and proapoptotic status; high level of IL-1β activates the proliferation of tumor cells, while TNF-α and IFN-γ have key roles in the pathogenesis of bone marrow failure. Below is an analysis of these endpoints.

1. TNFα

The pleiotropic cytokine named cachectin, previously known as Tumor Necrosis Factor alpha (TNF-α), and the molecule lymphotoxin alpha (LFT-α), formerly known as Tumor Necrosis Factor beta (TNF-β), are polypeptide hormones, members of the TNF superfamily that consists of approximately 30 related molecules encoded in MHC Class III region. The *ex novo* synthesized proTNF-α is placed on the plasma membrane and cleaved into a soluble form of 17 kilodaltons (TNFs) TNF-α converting enzyme (TACE or ADAM-17). TNF is mainly produced by macrophages, and by a broad variety of other tissues including lymphoid cells, mast cells, endothelial cells, fibroblasts and neuronal tissue. TNF-α and LFT-α bind the same two receptors on the cell surface, the 55 kDa (TNF-R Type I p55) and 75 kDa (TNF-R Type II p75). The signal transduction is mediated primarily by p55TNF-R, while the main action of p75 is the presentation of the TNF molecule to p55, with the function of "ligand passing" on the cellular membrane.

TNF-R1 is constitutively expressed in most tissues, whereas expression of TNF-R2 is highly regulated and is typically found in cells of the immune system. In most cells, TNF-R1 appears to be the key mediator of TNF signalling, whereas in the lymphoid system TNF-R2 seems to play a major role. Generally, the importance of TNFR2 is likely to be underestimated, because this receptor can only be fully activated by membrane-bound TNF (memTNF), but not by soluble TNF (sTNF).

TNF is considered to be a key effector in the development of septic shock. Whereas high concentrations of TNF induce shock-like symptoms, the prolonged exposure to low TNF concentrations can result in a wasting syndrome, which is cachexia. Owing to its strong proinflammatory and immunostimulatory activities, TNF is an important mediator of progression in many diseases (Wajant et al. 2003); in fact, the default pathway of TNF-α signaling is the induction of genes involved in inflammation and cell survival (Du et al.

2014). When TNF-α binds its receptor TNF-R1, TRADD protein (TNFR1-associated death domain protein) is recruited; downstream of TRADD, two signaling complexes are formed: 1) the plasma membrane-bound complex (complex I) consisting of TNFR1, TRADD, the receptor interacting protein (RIP), and TNF receptor-associated factor 2 (TRAF2), leading to rapid activation of NF-κB and the mitogen-activated protein kinases (MAPK) pathways, and 2) the cytoplasmic complex (complex II) containing TRADD, RIP, FAS-associated death domain protein (FADD), and caspase 8, essential for TNFα-induced apoptosis through a caspase cascade. In most cells the combination of the above signaling pathways determines the diverse biological activities of TNFα, including cell growth, development, oncogenesis, inflammation, stress-induced signaling, and cell death (Shen and Pervaiz 2006).

RIP-NF-κB axis acts as suppressor of intracellular ROS formation in response to TNF-α. This is possible due to antioxidant enzymes/proteins that are under NF-κB regulation; this is the case, e.g., for manganese superoxide dismutase (MnSOD), a mitochondrial activity that removes $^{\bullet}O_2^-$ derived from the mitochondrial respiratory chain, involved in the resistance to TNF-α-induced apoptosis. Another example is represented by ferritin heavy chain (FHC), the primary iron storage factor highly capable of suppressing ROS accumulation through iron sequestration, preventing sustained c-Jun N-terminal kinase (JNK) activation and TNF-α–triggered apoptosis.

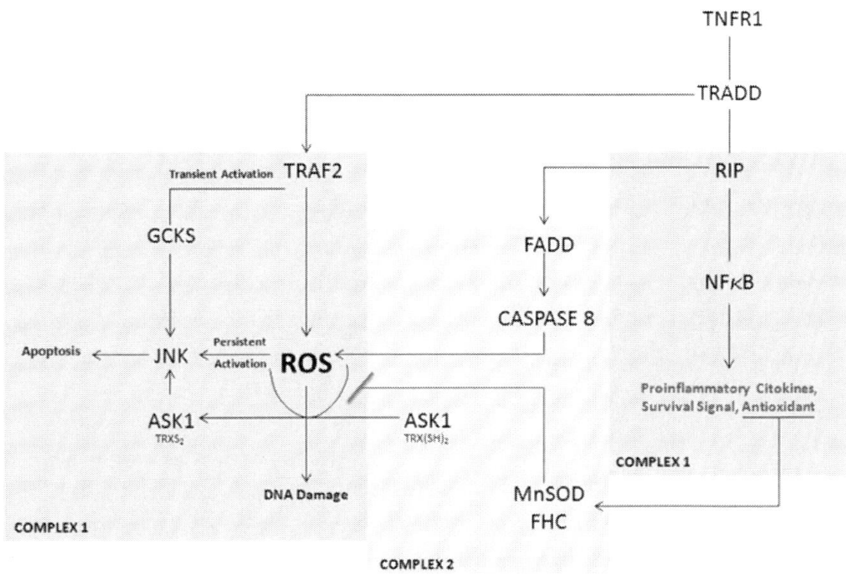

Figure 1. Major signal transduction pathways of TNF-α.

TNFR1= Tumor Necrosis Factor Receptor; TRADD = Tumor necrosis factor receptor type 1-associated DEATH domain protein; RIP = Receptor-interacting serine/threonine-protein kinase 1; NF-κB = Nuclear Factor Kappa-light-chain-Enhancer of activated B cells; FADD = Fas-Associated protein with Death Domain; TRAF2 = TNF receptor-associated factor 2; GCK_S = Germinal Center Kinases; JNK = c-Jun N-terminal kinase; ASK1 = Apoptosis Signal-regulated Kinase-1; TRXS2 = oxidized Thioredoxin; $TRX(SH)_2$ = reduced Thioredoxin; MnSOD = Manganous SuperOxide Dismutase; FHC = *Ferritin* Heavy Chain.

JNK activation by TNF can occur by caspase-dependent (GCKs or germinal center kinase) and/or caspase independent pathways based on ROS-dependent activation; in particular, transient activation of JNK is mediated by caspase-dependent way, while the delayed and persistent activation of JNK is mediated by ROS. JNK signalling is a pro-apoptotic signal.

Moreover, there is an important relationship between thioredoxin (Trx) and ASK1 (apoptosis signal-regulating kinase 1) an important JNK activator. In fact thioredoxin in its reduced form is capable of binding to ASK1 and blocking its kinase activity. Conversely, in presence of ROS, the oxidized thioredoxin dissociates from ASK1 that can activate JNK (Shen et al. 2006)

Complex II (RIP, FADD and caspase 8), essential for TNF-α-induced apoptosis through a caspase cascade, leads to the accumulation of intracellular ROS and mitochondrial pathway activation. Therefore, ROS production is both part of the activation of Complex I (at inflammatory sites, it causes DNA damage that can to lead to cellular transformation), and part of Complex II (Shen et al. 2006; Wajant et al. 2003).

TNF-α, as IL-1β and IFN-γ, increase both intracellular and extracellular ROS production in a time- and dose-dependent manner (Yang et al. 2007). Figure 1 highlights the pivotal role of ROS, fine regulators of homeostasis between apoptotic and survival signaling. Through these pathways, survival/inflammatory signaling and inducing death, TNF-α is involved in chronic inflammatory conditions correlated to pathologies as: diabetes, cancer, cardiovascular disease, septic shock and even anemia, leukemia and Fanconi Anemia.

1.1. TNFα in Fanconi Anemia

Increased TNF-α levels are often found in serum, plasma, cells or in bone marrow of FA patients (Schultz et al. 1993). Consequently, changes are observerd in the Bone Marrow microenvironment, deregulating cellular homeostasis and producing excess apoptosis in HSC compartment: hence the Bone Marrow Failure (BMF) found in FA children. Moreover, BMF would be responsible for the decreased production of B and NK cells, which represent the first defense against cancer (Castello et al. 1998; Myers et al. 2011).

Recently it has been suggested that FA patients show a decreased proportion of T CD8(+) and NK CD56(dim)CD16(+) cells when compared to healthy controls as well as an imbalance of the subsets NK lymphocytes. Data suggest that FA patients might have a defective cytotoxic response due to the lower number of cytotoxic cells as well as impairment in the differentiation process of the NK cells subsets which may be directly related to impairment of the immune surveillance observed in these patients (Justo et al. 2014).

Major antioxidant defense genes are down-regulated in the BM cells of FA patients; this down-regulation seems to be selectively associated with increased oxidative DNA damage in the promoters of these antioxidant defense genes. Loss of FA protein(s) makes DNA susceptible to ROS attack (Du et al. 2014; Pagano et al. 2013).

Figure 2 shows the main signal transduced in FA patients that induces HSC death and DNA damage through a mechanism of persistent activation. Persistent induction of this pathway promotes leukemic transformation of FA hematopoietic stem/progenitor cells.

ROS are implicated as chemical mediators in normal hematopoiesis; their level controls HSC self-renewal and proliferation. In particular, OS activates FoxO genes (forkhead O (FoxO) family of transcription factors) whose target genes are involved in cell cycle arrest, stress resistance, differentiation, apoptosis, and metabolism.

Several recent studies indicate that FoxO-dependent signaling is required for long-term regenerative potential of the HSC compartment through regulation of HSC response to physiologic OS, quiescence, and survival. In particular, Fancd2 was required for nuclear retention of ca-FOXO3a (active form of FOXO3a) and for maintaining hematopoietic repopulation of the HSCs (Li et al.. 2014).

This mechanism may play an important role in the maintenance and integrity of stem cell compartments in a broad spectrum of tissues (Tothova et al. 2007; Carter et al. 2007). In particular, elevated ROS levels reduce HSC self-renewal and inhibit hematopoiesis (Zhou et al. 2013). Moreover, oxidative DNA damage activates the expression of cell cycle inhibitors in a HSC-specific manner, leading to the premature senescence among HSCs, and ultimately to the loss of stem cell function (Yahata et al. 2011). According to a hypothetic model, ROS mediate the induction of HSC senescence under various pathophysiological conditions. Increased ROS levels are produced by mitochondria and/or NADPH oxidases (NOXs) in HSCs when HSCs are exposed to stress, a high oxygen concentration, persistent activation of TNF-α signal, or after deletion of *ATM*, *FoxOs*, *Bmi1*, *TSC1* and *HIF1α* (Tothova et al. 2007).

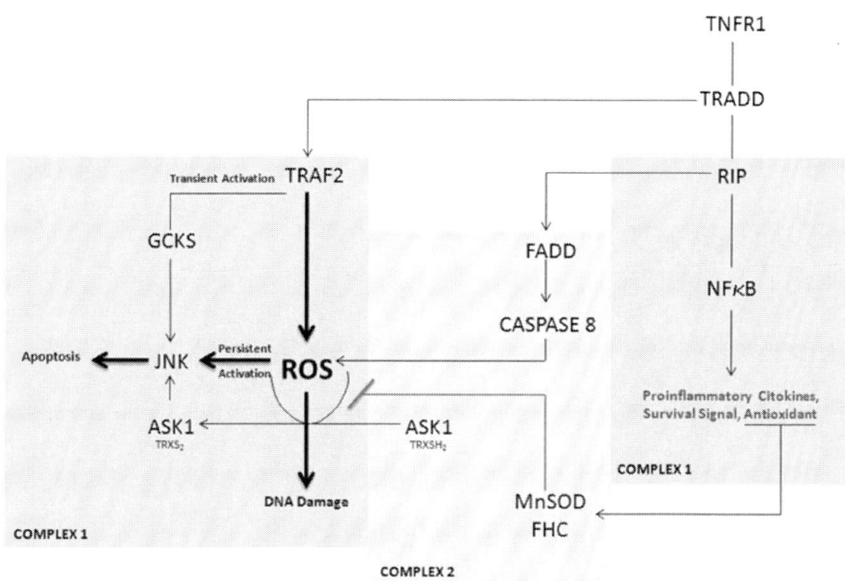

Figure 2. Major signal transduction pathways of TNF-α in Fanconi Anemia.

TNF-α-induced senescence correlates with the accumulation of ROS and oxidative DNA damage. Examination of senescence makers, such as senescence-associated beta-galactosidase, HP1-gamma, p53 and p16(INK4A) shows that TNFalpha induces premature senescence in bone marrow HSCs and progenitor cells as well as other tissues of Fancc-/-

mice. Pretreatment of TNF-α-injected Fancc-/- mice with a ROS scavenger significantly reduces oxidative base damage, DNA strand breaks and senescence (Zhang et al. 2007). In such a background of deficient hematopoiesis, FA BM cells experience a strong selective pressure toward clonal evolution, probably favored by the constitutive chromosomal instability (Bagby et al. 2007; Soulier 2011).

Acute leukemias are malignant clonal disorders of hematopoietic precursor cells, in which leukemic stem cells acquire mutations that confer self-renewal capabilities, altered hematopoietic differentiation, and increased proliferative capacity. (Lyu et al. 2008) ROS modulate the expression of redox-sensitive transcription factors, enzymes, oncogenes (Bcr/Abl, Ras and c-myc) and other downstream effectors. Ras-induced ROS, for example, promote growth factor-independent proliferation in human $CD34^+$ hematopoietic progenitor cells. ROS stress causes Double Strand Breaks (DSBs) and altered repair, leading to genomic changes and carcinogenesis. Several cytokines and growth factors were found to induce ROS production, which may alter the balance of expression of different genes. G-CSF and IL-3, in particular, stimulate ROS increase that can protect leukemic cells from apoptosis; in fact, elevated ROS levels trigger apoptosis in NK cells, CD4+ T cells and CD8+ T cells, whereas AML cells use ROS signaling as a strategy to evade cell-mediated immunity. This effect can be counteracted by antioxidants (Zhou et al. 2013).

2. IFN-γ

Another late mediator of the stem cell damage, Interferon-gamma (IFN-γ) is overexpressed in bone marrow lymphocytes from FA patients, together with TNF-α, IFN-γ is a non-genotoxic immunoregulatory lymphokine that can enhance both the innate and adaptive immune response against pathogens and tumors. It has the ability to maintain immune homeostasis; because of its high therapeutic index it has been used to treat a number of malignancies (Dufour et al. 2003; Lin et al. 2013).

Binding of IFN-γ to its receptor triggers the janus kinase (JAK/Signal Transducer) and Activator of Transcription (STAT) pathway. STAT1 is phosphorylated, dimerizes and then translocates into the nucleus to initiate the transcription or repression of target genes (Lin et al. 2013).

The ability of Stat1 to activate or repress gene transcription depends on the presence of other transcription factors binding to the promoter elements. Stat1 interaction with other factors such as IRF-9, upstream stimulatory factor-1, specificity protein-1, and heat shock factor-1 have been reported and are likely to influence specificity of DNA binding and transactivator ability. IFN-γ signaling also produces Stat1:Stat1:IFN regulatory factor (IRF)-9 and Stat1:Stat2:IRF-9 [IFN-stimulated gene factor 3 (ISGF3)] complexes. IFN-γ induced STAT1 phosphorylation in normal and in FANC-C B cells.

The major sources of IFN-γ are NK cells, T cells and NKT cells; its receptor is expressed ubiquitously on almost all cell types (Schroder et al. 2004).

2.1. IFN-γ in Fanconi Anemia

Recently, increasing research is focused on the impact of the pro-inflammatory cytokines on hematopoiesis, since HSC output is altered in response to these signals during infection, injury or radiation/chemotherapy. In particular, IFN-γ strongly diminishes the expansion of HSCs in the primary liquid culture of purified human $CD34^{+}CD38^{-}$ HSCs, negatively affects proliferation, increases apoptosis, and decreases self-renewing cell division due to its ability to inhibit STAT5 phosphorylation, which is an important positive regulator for HSC self-renewal (de Bruin et al. 2013; 2014).

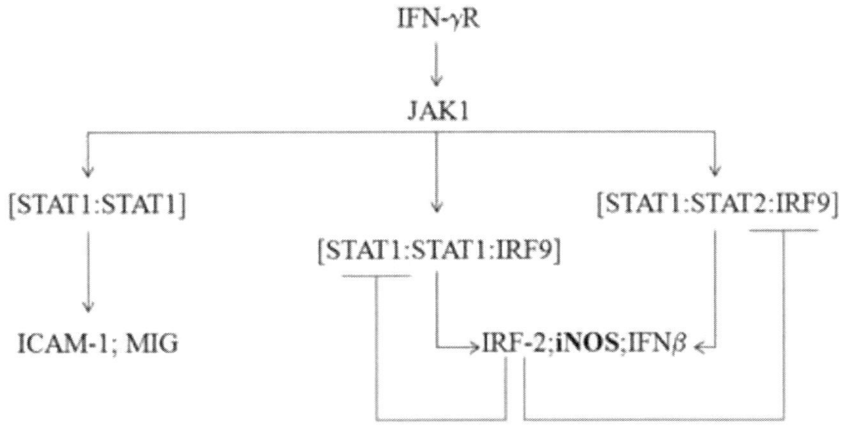

Figure 3. Major signals transduction pathways of IFN-γ in normal and FANC-C, B cells. IFN-γR = Interferon gamma Receptor; JAK1 = Janus kinase 1; STAT1 = Signal Transducer and Activator of Transcription 1; STAT2 = Signal Transducer and Activator of Transcription 2; IRF9 = Interferon Regulatory Factor 9; IRF 2 = Interferon Regulatory Factor 2; iNOS = Inducibile Nitric Oxide Synthase; IFN-β= Interferon beta; ICAM-1 = Intercellular Adhesion Molecule 1; MIG = Monokine Induced IFN-γ.

Several recent studies highlighted the role of IFNs as a major regulator of HSC cell cycle. Both IFN-α and IFN-γ stimulate HSC proliferation in mice and HSCs from IFN-γ-deficient mice are more quiescent and engraft better in transplantation assays than those from wild-type animals. Conversely, overactivation of IFN pathway impairs HSC functions and decreases their engraftment, suggesting a link between chronic infections and HSC exhaustion (Rossi et al. 2012). It has been demonstrated that stem cells ($CD34^{+}$) from adult, normal and FANC-C bone marrow express Fas constitutively and confirmed that IFN-γ augments Fas expression and primes cells for apoptosis after Fas ligation. Therefore, IFN-γ mediates hematopoietic suppression through the Fas pathway and FANC-C cells have hypersensitivity to IFN-γ. (Rathbun et al. 1997) A mechanism for the hypersensitivity of FANC-C-deficient cells to IFN-γ is mediated by FANC-C physically interacting with heat shock protein 70 to inhibit PKR activation and apoptosis(Pang et al. 2001) Subsequent studies showed that FANC-A and FANC-G also participate in that complex. FANC-A progenitors were found to be hypersensitive to IFN-γ via an increase in apoptosis (Si et al. 2006). In particular, *in vitro* studies showed that treatment with IFN-γ result in higher ROS accumulation in FA-C B-lymphocyte cell lines and this accumulation activates the JNK and p38 MAPK pathways in

FA lymphoid cell lines of the C complementation group. P38 MAPK functions as a signaling mediator of apoptosis (Yafe et al. 2004).

3. Other Cytokines: IL1-β, TGF-β, IL-6

IL-1β acts in concert with anti-inflammatory cytokines; in particular, IL-1R antagonist (IL-1Ra) is able to ensure the correct development and outcome of the inflammation: imbalance in the IL-1β/IL-1Ra ratio is implicated in many human diseases. IL-1β cytokine is produced by activated macrophages. This cytokine is an important mediator of the inflammatory response and is involved in different cellular activities, including cell proliferation, differentiation and apoptosis. The IL-1β/IL-1Ra unbalance after activation of multiple TLRs depends on the onset of OS, because of enhanced ROS production and failure of the antioxidant systems. Increased ROS levels increase ATP externalization by monocytes, resulting in enhanced inflammasome activation and IL-1β secretion. Oxidative stress is responsible for the impaired production of IL-1Ra. IL-1Ra secretion and is compensated by exogenous antioxidants (Lavieri et al. 2014). Moreover ROS induce the inflammasome that promotes the maturation of the other inflammatory cytokines as IL-1β (Martinon et al. 2002). TLR engagement is also found in FANC-C e FANC-A cells; it induces overproduction of both TNF-α and IL-1β in a p38-dependent manner (Garbati et al. 2013). Moreover, in FA-A cells the PI3K (phosphoinositide 3-kinase) - Akt pathway is constitutively activated; it induces an overexpression of IL-1β through an NF-κB independent mechanism and this overproduction activates the proliferation of tumor cells (Ibañez et al. 2009). Exposure of FANC-C-deficient BM progenitors to IL-1β potently produces suppression of the multipotent progenitor cells expansion *in vitro* (Garbati et al. 2013). IL1-β serum levels of other Fanconi complementation groups are not increased.

The role of IL-6 is really controversial because there is a reduced expression in patients' BM cells (Rosselli et al. 1992; Stark et al. 1993) and in peripheral blood mononuclear cells (PBMCs) (Bagnara et al. 1993). This may, in part, account for hematopoietic disease progression in FA and for deregulation of cellular homeostasis. In contrast, IL-6 was significantly higher in FA mesenchymal stem cell (MSCs) in Lecourt studies (Lecourt el al 2010). These controversial data are probably due to the dual role of IL-6 that can be a pro- or anti-inflammatory cytokine. In particular, IL-6 role as an anti-inflammatory cytokine is mediated through its inhibitory effects on TNF-α and IL-1 and activation of IL-1ra and IL-10. Since in FA patients TNF-α is overexpressed, we can speculate that this depends on the proinflammatory role of IL-6. Thus, IL-6 would have a pro-inflammatory role in FA.

Conclusion

Absolute numbers of B cells and NK cells are frequently reduced in FA patients compared to controls, while the absolute number of T cells is within normal range. Therefore FA patients might have a defective cytotoxic response due to the lower number of cytotoxic cells as well as to an impairment in the differentiation process of the NK cells subsets, which may be directly related to the impairment of the immune surveillance observed in these patients

Oxidative stress is considered to be an important pathogenic factor in FA, and ROS, generated during inflammation, seem to play the crucial role. Pro-inflammatory cytokines, as TNF-α, IL-1-β, and IFN-γ, induce ROS in human cells; however, the underlying mechanisms remain poorly understood. TNF-α, IL-1β and IFN-γ increase both intracellular and extracellular ROS production in a time and dose-dependent manner.. Moreover ROS induce the inflammasome that promotes the maturation of the other inflammatory cytokines as IL-1β The presence of ROS and inflammatory microenvironment predispose FA cells to malignant transformation and bone marrow failure. This inhibits HSC self-renewal with a well-known consequence in FA patients.

References

Bagby, G. C. & Meyers, G. (2007). Bone marrow failure as a risk factor for clonal evolution: prospects for leukemia prevention. *Hematol. Am. Soc. Hematol. Educ. Program.,* 40-46.

Carter, M. E. & Brunet, A. (2007). FOXO transcription factors. *Curr. Biol., 17*, 113-114.

Castello, G., Gallo, C., Napolitano, M. & Ascierto, P. A. (1998). Immunological phenotype analysis of patients with Fanconi's anaemia and their family members. *Acta Haematologica, 100*, 39–43.

De Bruin, A. M., Demirel, Ö., Hooibrink, B., Brandts, C. H. & Nolte, M. A. (2013). Interferon-gamma impairs proliferation of hematopoietic stem cells in mice. *Blood, 121*, 3578-3585

De Bruin, A. M., Voermans, C. & Nolte, M. A. (2014). Impact of interferon-γ on hematopoiesis. *Blood, 124*, 2479-2486.

Du, W., Erden, O. & Pang, Q. TNF-α signaling in Fanconi anemia. (2014). *Blood Cells Mol. Dis., 52*, 2-11.

Dufour, C., Corcione, A., Svahn, J., Haupt, R., Poggi, V., Bekassy, A. N., Scime, R., Pistorio, A. & Pistoia, V. (2003). TNF-α and IFN-γ are overexpressed in the bone marrow of Fanconi anemia patients and TNF-α suppresses erythropoiesis in vitro. *Blood, 102*, 2053-2059.

Garbati, M. R., Hays, L. E., Keeble, W., Yates, J. E., Rathbun, R. K. & Bagby, G. C. (2013). FANCA and FANCC modulate TLR and p38 MAPK–dependent expression of IL-1b in macrophages. *Blood, 122*, 3197- 3205.

Ibáñez, A., Río, P., Casado, J. A., Bueren, J. A., Fernández-Luna, J. L. & Pipaón, C. (2009). Elevated levels of IL-1*β* in Fanconi anaemia group A patients due to a constitutively active phosphoinositide 3-kinase-Akt pathway are capable of promoting tumour cell proliferation. *Biochem. J., 422*, 161–170.

Justo, G. A., Bitencourt, M. A., Pasquini, R., Castelo-Branco, M. T., Almeida-Oliveira, A., Diamond, H. R., Rumjanek, V. M. (2014) Immune status of Fanconi anemia patients: decrease in T CD8 and CD56dim CD16+ NK lymphocytes. *Ann Hematol., 5,* 761-767.

Korthof, E. T., Svahn, J., Peffault de Latour, R., Terranova, P., Moins-Teisserenc, H., Socie, G., Soulier, J., Kok, M., Bredius, R. G. M., van Tol, M., Jol-van der Zijde, E. C. M., Pistorio, A., Corsolini, F., Parodi, A., Battaglia, F., Pistoia, V., Dufour, C. & Cappelli, E.

(2013). Immunological profile of Fanconi anemia: A multicentric retrospective analysis of 61 patients. *Am. J. Hematol., 88*, 472-476.

Lavieri, R., Piccioli, P., Carta, S., Delfino, L., Castellani, P. & Rubartelli, A. (2014). TLR Costimulation Causes Oxidative Stress with Unbalance of Proinflammatory and Anti-Inflammatory Cytokine Production. *J. Immunol., 192*, 5373-5381.

Lecourt, S., Vanneaux, V., Leblanc, T., Leroux, G., Ternaux, B., Benbunan, M., Chomienne, C., Baruchel, A., Marolleau, J. P., Gluckman, E., Socié, G., Soulier, J. & Larghero, J. (2010). Bone Marrow Microenvironment in Fanconi Anemia: A Prospective Functional Study in a Cohort of Fanconi Anemia Patients. *Stem Cells Dev., 19*, 203-208.

Li, X., Li, J., Wilson, A., Sipple, J., Schick, J. & Pang, Q. (2014). Fancd2 is required for nuclear retention of FOXO3a in hematopoietic stem cell maintenance. *J. Biol. Chem.*, pii: jbc.M114.619536. [Epub ahead of print]

Lin, F. C. & Young, H. A. (2013). The talented interferon-gamma. *Adv. Biosci. Biotechnol., 4*, 6-13.

Lyu, B. N., Ismailov, S. B., Ismailov, B. & Lyu, M. B. (2008). Mitochondrial concept of leukemogenesis: key role of oxygen-peroxide effects. *Theor. Biol. Med. Model., 5*, 23.

Martinon, F., Burns, K. & Tschopp, J. (2002). The inflammasome: a molecular platform triggering activation of inflammatory caspases and processing of proIL-beta. *Mol. Cell., 10*, 417–426.

Myers, K. C., Bleesing, J. J., Davies, S. M., Zhang, X., Martin, L. J., Mueller, R., Harris, R. E., Filipovich, A. H., Kovacic, M. B., Wells, S. I. & Mehta, P. A. (2011). Impaired immune function in children with Fanconi anaemia. *Br. J. Haematol., 154*, 234-240.

Pagano, G., Aiello Talamanca, A., Castello, G., d'Ischia, M., Pallardó, F. V., Petrović, S., Porto, B., Tiano, L. & Zatterale, A. (2013). Bone marrow cell transcripts from Fanconi anaemia patients reveal in vivo alterations in mitochondrial, redox and DNA repair pathways. *Eur. J. Haematol., 91*, 141-151.

Pagano, G., Aiello Talamanca, A., Castello, G., Cordero, M. D., d'Ischia, M., Gadaleta, M. N., Pallardó, F. V., Petrović, S., Tiano, L. & Zatterale, A. (2014). Current experience in testing mitochondrial nutrients in disorders featuring oxidative stress and mitochondrial dysfunction: Rational design of chemoprevention trials. *Int. J. Mol. Sci., 15*, 20169-20208.

Pang, Q., Keeble, W., Christianson, T. A., Faulkner, G. R. & Bagby, G. C. (2001). FANCC interacts with Hsp70 to protect hematopoietic cells from IFN-gamma/TNF-alpha-mediated cytotoxicity. *EMBO J., 20*, 4478-4489.

Pearl-Yafe, M., Halperin, D., Scheuerman, O. & Fabian, I. (2004). The p38 pathway partially mediates caspase-3 activation induced by reactive oxygen species in Fanconi anemia C cells. *Biochem. Pharmacol., 67*, 539-546.

Rathbun, R. K., Faulkner, G. R., Ostroski, M. H., Christianson, T. A., Hughes, G., Jones, G., Cahn, R., Maziarz, R., Royle, G., Keeble, W., Heinrich, M. C., Grompe, M., Tower, P. A. & Bagby, G. C. (1997). Inactivation of the Fanconi anemia group C gene augments interferon-gamma-induced apoptotic responses in hematopoietic cells. *Blood, 90*, 974-985.

Rosselli, F., Sanceau, J., Wietzerbin, J. & Moustacchi, E. (1992). Abnormal lymphokine production: a novel feature of the genetic disease Fanconi anemia. I. Involvement of interleukin-6. *Hum. Genet., 89*, 42–48.

Rossi, L., Salvestrini, V., Ferrari, D., Di Virgilio, F. & Lemoli, R. M. (2012). The sixth sense: hematopoietic stem cells detect danger through purinergic signaling. *Blood, 120*, 2365-2375.

Shen, H. M. & Pervaiz, S. (2006). TNF receptor superfamily-induced cell death: redox-dependent execution. *FASEB J., 10*, 1589-1598.

Schroder, K., Hertzog, P. J., Ravasi, T. & Hume, D. A. (2004). Interferon-γ: an overview of signals, mechanisms and functions. *J. Leukoc. Biol., 75*, 163-189.

Schultz, J. C. & Shahidi, N. T. (1993). Tumor necrosis factor-alpha overproduction in Fanconi's anemia. *Am. J. Hematol., 42*, 196–201.

Shen, H. M. & Pervaiz, S. (2006). TNF receptor superfamily-induced cell death: redox dependent Execution. *FASEB J., 20*, 1589-1598.

Si, Y., Ciccone, S., Yang, F. C., Yuan, J., Zeng, D., Chen, S., van de Vrugt, H. J., Critser, J., Arwert, F., Haneline, L. S. & Clapp, D. W. (2006). Continuous in vivo infusion of interferon-gamma (IFN-γ) enhances engraftment of syngeneic wild-type cells in *fanca*$^{-/-}$ and *Fancg*$^{-/-}$mice. *Blood, 108*, 4283-4287.

Soulier, J. (2011). Fanconi Anemia. Understanding and management of inherited bone marrow failure syndromes. *Hematology Am. Soc. Hematol. Educ. Program,* 492-497.

Stark, R., Andre, C., Thierry, D., Cherel, M., Galibert, F. & Gluckman, E. (1993). The expression of cytokine and cytokine receptor genes in long-term bone marrow culture in congenital and acquired bone marrow hypoplasias. *Br. J. Haematol., 83*, 560–566.

Tothova, Z. & Gilliland, D. G. *(2007). Foxo transcription factors and stem cell homeostasis: insight from the hemapoietic system. Cell Stem Cell, 1*, 140-152.

Wajant, H., Pfizenmaier, K. & Scheurich, P. (2003). Tumor necrosis factor signaling. *Cell Death Diff., 10*, 45–65.

Yahata, T., Takanashi, T., Muguruma, Y., Ibrahim, A. A., Matsuzawa, H., Uno, T., Sheng, Y., Onizuka, M., Ito, M., Kato, S. & Ando, K. (2011). Accumulation of oxidative DNA damage restricts the self-renewal capacity of human hematopoietic stem cells. *Blood, 118*, 2941-2950.

Yang, D., Elner, S. G., Bian, Z. M., Till, G. O., Petty, H. R. & Elner, V. M. (2007). Pro-inflammatory cytokines increase reactive oxygen species through mitochondria and nadph oxidase in cultured rpe cells. *Exp. Eye Res., 85*, 462-472.

Zhang, X., Sejas, D. P., Qiu, Y., Williams, D. A. & Pang, Q. (2007). Inflammatory ROS promote and cooperate with the Fanconi anemia mutation for hematopoietic senescence. *J. Cell Sci., Pt 9*, 1572-1583.

Zhou, F., Shen, Q. & Claret, F. X. (2013). Novel roles of reactive oxygen species in the pathogenesis of acute myeloid leukemia. *J. Leukoc.Biol., 94*, 423-429.

In: Fanconi Anemia and Oxidative Stress
Editor: Giovanni Pagano

ISBN: 978-1-63482-297-8

Chapter 8

Need for in Vivo Characterization of Oxidative Stress and Mitochondrial Dysfunction in Fanconi Anemia patients

***Giovanni Pagano*[*,1] *and Luca Tiano*[†,2]**
[1]Italian National Cancer Institute, CROM, Italy
[2]DISCO, Polytechnic University of Marche, Italy

Abstract

The relevance and pathogenetic roles of oxidative stress (OS) and mitochondrial dysfunction (MDF), unconfined to Fanconi Anemia (FA), are shared to a number of genetic and acquired diseases, ranging from mitochondrial diseases (MDs) to several acquired disorders. The key-roles of mitochondria in maintaining cellular redox balance and, then, the involvement of MDF in several pathogenetic events are discussed. Investigating the *in vivo* implications of MDF in FA, and in a number of related and unrelated disorders may open unprecedented avenues both in elucidating pathogenetic events and in prompting mitochondria-oriented clinical research and patient management.

Keywords: oxidative stress, mitochondria, mitochondrial dysfunction, mitochondrial diseases, mitochondrial nutrients

Introduction

Far from being a unique phenomenon, the occurrence of OS and MDF, as documented in FA (Pagano et al. 2013), has been reported in a number of genetic diseases (GDs), as summarized in Table 1. Moreover, energetic metabolism impairment and MDF have been

[*]I-83013 Mercogliano (AV), Italy; E-mail: gbpagano@tin.
[†]I-60131 Ancona, Italy; E-mail: l.tiano@univpm.it.

implicated in broad-ranging metabolic, degenerative, neoplastic, and age-related diseases. Mitochondrial diseases are characterized by defects in the mitochondrial respiratory chain that result in deficient ATP production. Nonetheless MDs are more frequent than it was previously thought: 1 in 10,000 individuals has mitochondrial disease, and mtDNA mutations are found as frequently as in 1 of every 200 samples of cord blood (Elliott et al. 2008; Chinnery et al. 2012). Although tissues with high energy demands (heart, brain, muscle) are primarily targeted, MDF can affect any organ triggering the development of systemic multifactorial diseases such as diabetes and cardiovascular diseases. Three classes of MDs are known: (1) primary MDs caused by mutations in mitochondrial genes; (2) disorders with mutations in nuclear genes involved in mitochondrial function; and (3) secondary disorders that arise from the accumulation of mitochondrial damage over time frequently involving neurodegenerative pathologies (Ciccone et al. 2013, Karunadharma et al. 2010; Hauser and Hastings 2013).

Moreover, in several inborn metabolic diseases, associated with defects in enzymes involved in the intermediate metabolism, the genetic nature of the disease is not evident since often there seems to be no correlation between the genotype and the phenotype due to mutations that do not compromise the expression of the proteins but eventually lead to unstable and short-lived products that might underlie mild to severe disorders (Olsen et al. 2013).

Some GDs display an involvement of OS/MDF in their phenotype, as it is the well-established case, e.g., for Down Syndrome (DS) and for Ataxia-telangiectasia (AT) (reviewed by Pagano et al. 2014a).

A number of other GDs were found to associate OS and MDF features in their molecular, cellular and/or clinical phenotypes. This is, by definition, the case for a set of GDs termed MDs, where the genetic defects were found associated with the mitochondrial genome (reviewed by Koene and Smeitink 2009; Gillis and Kaye 2002). Beyond MDs, a number of other GDs with defects located to nuclear genome have shown to combine both MDF and OS in their respective phenotypes. Examples of GDs displaying both OS and MDF were provided by FA and other cancer- and/or aging-related diseases, as AT, DS and Werner Syndrome (WS) (Pallardó et al. 2010). Moreover, as shown in Table 1, other GDs have been reported to display both OS and MDF phenotypic features, as in the cases, among others, for Xeroderma pigmentosum, Cockayne Syndrome, Adrenoleukodystrophy, and Friedreich Ataxia (FrA) (Rezvani et al. 2011; Terrill et al. 2013; Sparaco et al. 2009).

The typical changes in OS endpoints included excess ROS production and oxidative DNA damage, lipid peroxidation and decrease in reduced glutathione (GSH:GSSH ratio). Together, MDF endpoints were found including decreased mitochondrial membrane potential ($\Delta\Psi$), decreased Krebs cycle or oxidative phosphorylation (OXPHOS) activities, or decreased ATP production. One of the abnormalities detected in MDF consisted of decreased levels of coenzyme Q10 (CoQ10), a cofactor in the OXPHOS pathway, as reported for DS and for FrA (Littarru and Tiano 2010; Tiano and Busciglio 2011; Miles et al. 2007; Schöls et al. 2005; Sparaco et al. 2009). Defective CoQ10 levels were reported for several MDs and for a number of other OS/MDF-related disorders (Pagano et al. 2014a).

Table 1. Occurrence of OS and/or MDF in the phenotypes of some selected genetic diseases*

Ataxia-telangiectasia	↓ $\Delta\Psi$; mt aberrations; ↑ ROS production and oxidative DNA damage	Barzilai et al. 2002; Ambrose et al. 2007; Valentin-Vega et al. 2012
Xeroderma pigmentosum	↓ repair of oxidative DNA damage; ↑ lipid peroxidation; defective mt gene transcripts	Xia et al. 1994; Tanaka et al. 1998; Hayashi 2008; Rezvani et al. 2011;
Down Syndrome	SOD-1 overexpression; ↓ Krebs cycle activities, ↓ CoQ10 and mt aberrations	Tiano and Busciglio 2011; Tiano et al. 2012; Pallardó et al. 2006; Roat et al. 2007
Werner Syndrome	↑ multiple anomalies in OS endpoints; mt aberrations in murine WS model	Bohr et al. 2000; Von Kobbe et al. 2004; Pagano et al. 2005; Pallardó et al. 2010
Cockayne Syndrome	↑ ROS production and oxidative DNA damage; ↓ 8-oxo-Gua DNA glycosylase-1; ↑ mitophagy	Javeri et al. 2011; Pascucci et al. 2012; Scheibye-Knudsen et al. 2012;
Adrenoleukodystrophy	↓ GSH:GSSH ratio; ↑ lipid oxidation; ↓ total antioxidant defense; ↓ OXPHOS activities	Renjini et al. 2012; Terrill et al. 2013;
Friedreich Ataxia	↓ Complex I, II and III, aconitase, and CoQ10	Cooper et al. 2008; Sparaco et al. 2009

* **Abbreviations**: mt = mitochondrial; $\Delta\Psi$ = mitochondrial membrane potential; OXPHOS = oxidative phosphorylation; 8-oxo-Gua = 8-oxoguanine.

The causes of CoQ10 deficiency in patients with different GDs, not directly involved in CoQ10 synthesis but affecting other genes like APTX (aprataxin) and mtDNA (Rahman et al. 2012; Horvath et al. 2012) are not known. Common to these patients is mitochondrial energy deficiency and it has been suggested that this could impair mitochondrial import of CoQ proteins thereby causing CoQ10 deficiency (Yen et al. 2011).

Mitochondria As a Critical Site in Cell Physiology and a Potential Target in Several Genetic Diseases

Mitochondria are crucial to the cellular energy production that fuels growth and reproduction. Beside supplying energy in the form of adenosine triphosphate (ATP), mitochondria generate reactive oxygen species (ROS) and modulate the relative signaling cascade, buffer cytosolic calcium ions, and regulate the mitochondrial permeability transition pore (Wallace et al. 2010). As a result, mitochondria act as gate keepers in a process that can steer cellular prospect to survival, apoptosis or necrosis.

Given the essential role of these organelles in cellular bioenergetics, it is not surprising that MDF is a hallmark of several inborn and acquired diseases, as well as its direct oxidative implications that represent a common factor in the pathophysiological mechanism of these diseases. However, despite this association between disease and OS, a detailed knowledge of the sources and types of the involved ROS is often still lacking.

The metabolism of ROS has double-edged implications for cellular physiology. For long time ROS have been considered a necessary toxic by-product of the mitochondrial respiration. Their toxicity derives from their instability and hence their ability to readily oxidize biomolecules as lipids, amino acids in proteins and DNA bases thereby damaging their functions while promoting additional ROS/RNS production, cell dysfunction and death. Moreover, oxidative stress also affects mitochondrial dynamics, mitochondrial trafficking, and transcriptional activity.

Considering the central role of mitochondria beyond energy metabolism, it is quite evident that the mantainence of the oxidant/antioxidant equilibrium and prevention of oxidative stress of these compounds is required to promote cellular health.

However, in light of the complexity of the signaling network, today still in large part unknown, the efficacy of a "standard" antioxidant therapy to counteract MDF and OS could result a simplicistic option; this was the case, e.g., of using N-acetylcysteine or resveratrol in compensating FA-related OS (Zhang et al. 2014; Columbaro et al. 2014). On the other hand, a combination of mitochondria-targeted molecules, independently and beyond their antioxidant activity, might provide an optimization of the overall mitochondrial functionality more likely to provide an effective support to the physiology of such a complex cellular organelle. A line of clinical research has been addressed to compensating MDF by administration of CoQ10, and/or other two cofactors involved in mitochondrial functions, i.e., α-lipoic acid (ALA) and l-carnitine (CARN) or acyl-l-carnitine. These cofactors, CoQ10, ALA and CARN are collectively termed "mitochondrial nutrients" (MNs) and have been tested in an number or clinical trials focused on mitochondrial diseases (Ogasahara et al. 1986; Scarlato et al. 1991; Barbiroli et al. 1997; Abe et al. 1999; Rodriguez et al. 2007; Glover et al. 2010; reviewed by

Pfeffer and Chinnery 2013; Pagano et al. 2014b). Beyond their use in MDs, MNs have been tested in clinical trials focused on an extensive number of genetic, aging-related and neurodegenerative diseases (reviewed by Pagano et al. 2014b; Folkers and Simonsen 1995; Ellaway et al. 2001; El-Beshlawy et al. 2004; Torrioli et al. 2008; Martins et al. 2009; Karimi et al. 2010; Geier et al. 2011; Spurney et a. 2011).

Mitochondrial Nutrients

Mitochondrial nutrients more than simply scavenging excessive mitochondrial ROS may have a more global impact on boosting mitochondrial OXPHOS activity. In particular, the proposed combination beside their powerful antioxidant activity, is able to support key processes in mitochondrial bioenergetics at the electron transport chain (ETC) level, promoting the activity of Krebs cycle dehydrogenases and fatty acid oxidation (FAO). The nature of these mitochondrial cofactors is also peculiar being in different proportions both endogenous and introduced through the diet. Their endogenous production has been shown to be limited in different diseased conditions and therapeutic supplementation represent an obvious means of re-equilibrating these levels (Miles et al. 2007; Tiano et al. 2012). In particular, for Coenzyme Q10 it is important to take into consideration that the physiological levels within the mitochondrial membranes are in the same concentration-range of the Michaelis–Menten constant (Km) of NADH oxidation at the respiratory complexes, so that even a slight decrease in CoQ10 content could result in a remarkable decrease in the ETC functionality (Estornell et al. 1992). On the other hand, this also means that the physiological concentration is not saturating and even a small increase in the CoQ10 content of mitochondrial membranes could lead to an increased respiratory rate. The control exerted by CoQ concentration over mitochondrial respiration could be of particular interest in situations of decreased CoQ10 levels and/or increased Km values for the quinone in some pathological states. This has been proposed as the mechanism of action by which exogenous CoQ10 ameliorates bioenergetic impairement in some mitochondrial myopathies and in cardiomyopathies (Mortensen and Mortensen 2014; Quinzii et al. 2008). The synergistic interaction of the different MNs in the combination would be able to efficiently fuel the mitochondrial ETC optimizing its functionality, as discussed by us previously (Pagano et al. 2014b). Beside their strictly bioenergetic effects, both CoQ10 and ALA are able to carry out their antioxidant activity synergistically with other cellular antioxidants. However, their scavenging activity has been shown not to curb the mitochondrial ROS signaling. On the contrary, both compounds (McCarty et al. 2009; Lee et al. 2012) as well as CARN (Sue et al. 2014) have been shown to induce AMPK-mediated PPARα activity promoting mitochondrial biogenesis.

Toward Mitochondria-Oriented Clinical Trials

Ascertaining an *in vivo* occurrence of MDF in FA, from the evidence observed in cell lines up to *in vivo* evidence from freshly drawn patients' cells and plasma, would be of paramount relevance. Solving this so far unexplored subject would both shed light into FA pathogenesis and in the prospect of planning mitochondria-oriented clinical interventions.

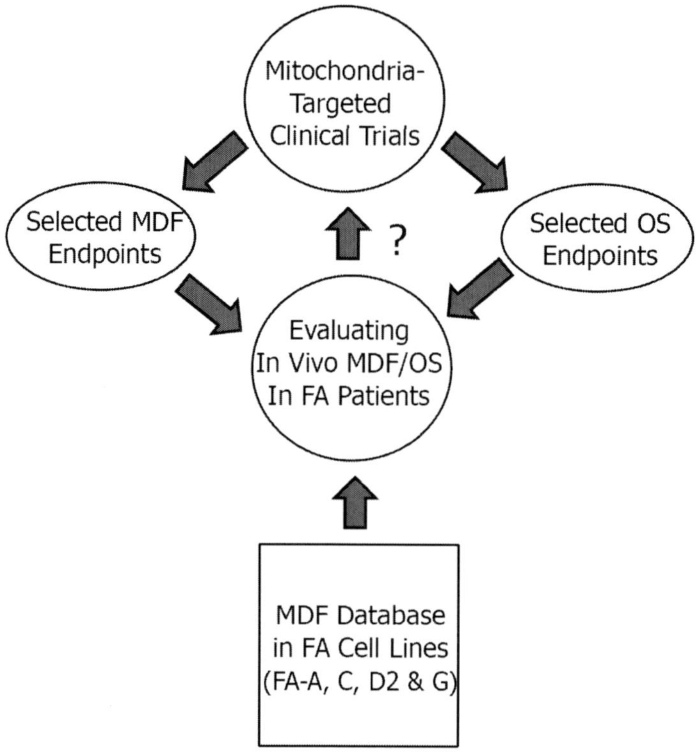

Figure 1. Prospect research development from the current database of OS/MDF endpoints in FA cells lines to *in vivo* evaluation of some selected OS/MDF endpoints, with subsequent design of evidence-based mitochondria-targeted clinical trials.

Based on the up-to-date evidence for MDF in FA cell lines, selected MDF and OS endpoints should be tested in blood cells and in biological fluids of FA patients, as depicted in Figure 1. If MDF is confirmed in this analytical step, one would acquire the evidence for defects in some defined endpoints of mitochondrial function such as, e.g., CoQ10 deficiency and/or other MDF endpoints. These findings are expected, as far as one may hardly figure out that MDF-related changes were found in FA cell lines and not in freshly drawn blood cells and/or biological fluids from FA patients. As for OS endpoints, the available database both points to *in vitro* and *in vivo* abnormalities.

The OS/MDF endpoints, found as abnormal in the preliminary step, shall be the grounds for choosing the best appropriate supplements to be administered to patients (Figure 1). The follow-up period will be of paramount relevance; indeed, short administration regimens may offer hints for efficacy and safety, yet they may conceal long-run effects. In a disease with multi-year-long progression, as FA, expecting the desired outcomes may be a quite elusive task. Thus, follow-up ought to be designed for a reasonably long duration, allowing monitoring the outcomes of multi-year-long administration, in terms of compensating the OS/MDF endpoints found affected in the analytical evaluation step. The ultimate goal, by compensating OS/MDF endpoints, shall be the delay or amelioration of disease progression.

Unconfined to FA clinical research, materializing the above prospect may have dramatic consequences in the success of to-be clinical trials for other OS/MDF-related disorders, trespassing from "hopes" toward a rational approach planning properly targeted interventions, then with evidence-based expectations.

References

Abe, K., Matsuo, Y., Kadekawa, J., Inoue, S. & Yanagihara, T. (1999). Effect of coenzyme Q10 in patients with mitochondrial myopathy, encephalopathy, lactic acidosis, and stroke-like episodes (MELAS): evaluation by noninvasive tissue oximetry. *J. Neurol. Sci., 162*, 65-68.

Ambrose, M., Goldstine, J. V. & Gatti, R. A. (2007). Intrinsic mitochondrial dysfunction in ATM-deficient lymphoblastoid cells. *Hum. Mol. Genet., 16,* 2154-2164.

Barbiroli, B., Frassineti, C., Martinelli, P., Iotti, S., Lodi, R., Cortelli, P. & Montagna, P. (1997). Coenzyme Q10 improves mitochondrial respiration in patients with mitochondrial cytopathies. An in vivo study on brain and skeletal muscle by phosphorous magnetic resonance spectroscopy. *Cell. Mol. Biol. (Noisy-le-grand), 43*, 741-749.

Barzilai, A., Rotman, G. & Shiloh, Y. (2002). ATM deficiency and oxidative stress: a new dimension of defective response to DNA damage. *DNA Repair (Amst), 1*, 3-25.

Bohr, V. A., Cooper, M., Orren, D., Machwe, A., Piotrowski, J., Sommers, J., Karmakar, P. & Brosh, R. (2000). Werner syndrome protein: biochemical properties and functional interactions. *Exp. Gerontol., 35*, 695-702.

Chinnery, P. F., Elliott, H. R., Hudson, G., Samuels, D. C. & Relton, C. L. (2012). Epigenetics, epidemiology and mitochondrial DNA diseases. *Int. J. Epidemiol., 41*, 177-187.

Ciccone, S., Maiani, E., Bellusci, G., Diederich, M. & Gonfloni, S. (2013) Parkinson's disease: a complex interplay of mitochondrial DNA alterations and oxidative stress. *Int. J. Mol. Sci., 14*, 2388-2409.

Columbaro, M., Ravera, S., Capanni, C., Panfoli, I., Cuccarolo, P., Stroppiana, G., Degan, P. & Cappelli, E. (2014). Treatment of FANCA cells with resveratrol and N-acetylcysteine: a comparative study. *PLoS One, 9*, 104857.

Cooper, J. M., Korlipara, L. V., Hart, P. E., Bradley, J. L. & Schapira, A. H. (2008). Coenzyme Q10 and vitamin E deficiency in Friedreich's ataxia: predictor of efficacy of vitamin E and coenzyme Q10 therapy. *Eur. J. Neurol., 15*, 1371-1379.

El-Beshlawy, A., Ragab, L., Fattah, A. A., Ibrahim, I. Y., Hamdy, M., Makhlouf, A., Aoun, E., Hoffbrand, V. & Taher, A. (2004). Improvement of cardiac function in thalassemia major treated with L-carnitine. *Acta Haematol., 111*, 143-148.

Ellaway, C. J., Peat, J., Williams, K., Leonard, H. & Christodoulou, J. (2001). Medium-term open label trial of L-carnitine in Rett syndrome. *Brain Dev., 23 (Suppl 1)*, S85-89.

Elliott, H. R., Samuels, D. C., Eden, J. A., Relton, C. L. & Chinnery, P. F. (2008). Pathogenic mitochondrial DNA mutations are common in the general population. *Am. J. Hum. Genet., 83*, 254-260.

Estornell, E., Fato, R., Castelluccio, C., Cavazzoni, M., Parenti Castelli, G. & Lenaz, G. (1992). Saturation kinetics of coenzyme Q in NADH and succinate oxidation in beef heart mitochondria. *FEBS Lett., 311*, 107-109.

Folkers, K. & Simonsen, R. (1995). Two successful double-blind trials with coenzyme Q10 (vitamin Q10) on muscular dystrophies and neurogenic atrophies. *Biochim. Biophys. Acta, 1271*, 281-286.

Geier, D. A., Kern, J. K., Davis, G., King, P. G., Adams, J. B., Young, J. L. & Geier, M. R. (2011). A prospective double-blind, randomized clinical trial of levocarnitine to treat autism spectrum disorders. *Med. Sci. Monit., 17*, PI15-23.

Gillis, L. & Kaye, E. (2002). Diagnosis and management of mitochondrial diseases. *Pediatr. Clin. North Am., 49*, 203-219.

Glover, E. I., Martin, J., Maher. A., Thornhill, R. E., Moran. G. R. & Tarnopolsky, M. A. (2010). A randomized trial of coenzyme Q10 in mitochondrial disorders. *Muscle Nerve, 42*, 739-748.

Hauser, D. N. & Hastings, T. G. (2013). Mitochondrial dysfunction and oxidative stress in Parkinson's disease and monogenic parkinsonism. *Neurobiol. Dis., 51*, 35-42.

Hayashi, M. (2008). Roles of oxidative stress in xeroderma pigmentosum. *Adv. Exp. Med. Biol., 637*, 120-127.

Horvath, R. (2012). Update on clinical aspects and treatment of selected vitamin-responsive disorders II (riboflavin and CoQ 10). *J. Inherit. Metab. Dis., 35*, 679-687.

Javeri, A., Lyons, J. G., Huang, X. X. & Halliday, G. M. (2011). Downregulation of Cockayne syndrome B protein reduces human 8-oxoguanine DNA glycosylase-1 expression and repair of UV radiation-induced 8-oxo-7,8-dihydro-2'-deoxyguanine. *Cancer Sci., 102*, 1651-1658.

Karimi, M., Mohammadi, F., Behmanesh, F., Samani, S. M., Borzouee, M., Amoozgar, H. & Haghpanah, S. (2010). Effect of combination therapy of hydroxyurea with l-carnitine and magnesium chloride on hematologic parameters and cardiac function of patients with beta-thalassemia intermedia. *Eur. J. Haematol., 84*, 52-58.

Karunadharma, P. P., Nordgaard, C. L., Olsen, T. W. & Ferrington, D. A. (2010). Mitochondrial DNA damage as a potential mechanism for age-related macular degeneration. *Invest. Ophthalmol. Vis. Sci., 51*, 5470-5479.

Koene, S. & Smeitink, J. (2009). Mitochondrial medicine: entering the era of treatment. *J. Intern. Med., 265*, 193-209.

Kumari, U., Jun, W., Bay, B. & Lyakhovich, A. (2014). Evidence of mitochondrial dysfunction and impaired ROS detoxifying machinery in Fanconi anemia cells. *Oncogene, 33*, 165–172.

Lee, S. K., Lee, J. O., Kim, J. H., Kim, N., You, G. Y., Moon, J. W., Sha, J., Kim, S. J., Lee, Y. W., Kang, H. J., Park, S. H. & Kim, H. S. (2012). Coenzyme Q10 increases the fatty acid oxidation through AMPK-mediated PPARα induction in 3T3-L1 preadipocytes. *Cell Signal., 24*, 2329-2336.

Littarru, G.P. & Tiano, L. (2010). Clinical aspects of coenzyme Q10: an update. *Nutrition, 26*, 250-254.

Martins, V. D., Manfredini, V., Peralba, M. C., Benfato, M. S. (2009). Alpha-lipoic acid modifies oxidative stress parameters in sickle cell trait subjects and sickle cell patients. *Clin. Nutr., 28*, 192-197.

McCarty, M. F., Barroso-Aranda, J. & Contreras, F. (2009). The "rejuvenatory" impact of lipoic acid on mitochondrial function in aging rats may reflect induction and activation of PPAR-gamma coactivator-1alpha. *Med. Hypotheses, 72*, 29-33.

Miles, M. V., Patterson, B. J., Chalfonte-Evans, M. L., Horn, P. S., Hickey, F. J., Schapiro, M. B., Steele, P. E., Tang, P. H. & Hotze, S. L. (2007). Coenzyme Q10 (ubiquinol-10) supplementation improves oxidative imbalance in children with trisomy 21. *Pediatr. Neurol., 37*, 398-403.

Mortensen, S. A. & Mortensen, A. L. (2014). The mitochondria in heart failure: a target for coenzyme Q10 therapy? *Clin. Pharmacol. Ther., 96*, 645-647.

Mukhopadhyay, S., Leung, K., Hicks, M., Hastings, P., Youssoufian, H. & Plon, S. (2006). Defective mitochondrial peroxiredoxin-3 results in sensitivity to oxidative stress in Fanconi anemia. *J. Cell Biol., 175*, 225-235.

Ogasahara, S., Nishikawa, Y., Yorifuji, S., Soga, F., Nakamura, Y., Takahashi, M., Hashimoto, S., Kono, N. & Tarui, S. (1986). Treatment of Kearns-Sayre syndrome with coenzyme Q10. *Neurology, 36*, 45-53.

Olsen, R.. K., Cornelius, N. & Gregersen, N. (2013) Genetic and cellular modifiers of oxidative stress: what can we learn from fatty acid oxidation defects? *Mol. Genet. Metab., 110 Suppl.,* S31-39.

Pagano, G., Zatterale, A., Degan, P., d'Ischia, M., Kelly, F. J., Pallardó, F. V. & Kodama, S. (2005). Multiple involvement of oxidative stress in Werner syndrome phenotype. *Biogerontology, 6*, 233-243.

Pagano, G., Aiello Talamanca, A., Castello, G., d'Ischia, M., Pallardó, F. V., Petrović, S., Porto, B., Tiano, L. & Zatterale, A. (2013). From clinical description to in vitro and animal studies, and backwards to patients: Oxidative stress and mitochondrial dysfunction in Fanconi anemia. *Free Radic. Biol. Med., 58*, 118-125.

Pagano, G., Talamanca, A. A., Castello, G., Cordero, M. D., d'Ischia, M., Gadaleta, M. N., Pallardó, F. V., Petrović, S., Tiano, L. & Zatterale, A. (2014a). Oxidative stress and mitochondrial dysfunction across broad-ranging pathologies: toward mitochondria-targeted clinical strategies. *Oxid. Med. Cell. Longev., 2014*, 541230.

Pagano, G., Talamanca, A. A., Castello, G., Cordero, M. D., d'Ischia, M., Gadaleta, M. N., Pallardó, F. V., Petrović, S., Tiano, L. & Zatterale, A. (2014b). Current experience in testing mitochondrial nutrients in disorders featuring oxidative stress and mitochondrial dysfunction: rational design of chemoprevention trials. *Int. J. Mol. Sci., 15*, 20169-20208.

Pallardó, F. V., Degan, P., d'Ischia, M., Kelly, F. J., Zatterale, A., Calzone, R., Castello, G., Fernandez-Delgado, R., Dunster, C., Lloret, A., Manini, P., Pisanti, M. A., Vuttariello, E. & Pagano, G. (2006). Multiple evidence for an early age pro-oxidant state in Down Syndrome patients. *Biogerontology, 7*, 211-220.

Pallardó, F. V., Lloret, A., Lebel, M., d'Ischia, M., Cogger, V. C., Le Couteur D. G., Gadaleta, M. N., Castello, G. & Pagano, G. (2010). Mitochondrial dysfunction in some oxidative stress-related genetic diseases: Ataxia-Telangiectasia, Down Syndrome, Fanconi Anaemia and Werner Syndrome. *Biogerontology, 11*, 401-419.

Pascucci, B., Lemma, T., Iorio, E., Giovannini, S., Vaz, B., Iavarone, I., Calcagnile, A., Narciso, L., Degan, P., Podo, F., Roginskya, V., Janjic, B.M., Van Houten, B., Stefanini, M., Dogliotti, E. & D'Errico, M. (2012). An altered redox balance mediates the hypersensitivity of Cockayne syndrome primary fibroblasts to oxidative stress. *Aging Cell, 11*, 520-529.

Pfeffer, G. & Chinnery, P. F. (2013). Diagnosis and treatment of mitochondrial myopathies. *Ann. Med., 45*, 4-16.

Quinzii, C. M., López, L. C., Naini, A., Di Mauro, S. & Hirano, M. (2008). Human CoQ10 deficiencies. *Biofactors, 32*, 113-8.

Rahman, S., Clarke, C. F. & Hirano, M. (2012). 176th ENMC International Workshop: diagnosis and treatment of coenzyme Q_{10} deficiency. *Neuromuscul Disord., 22*, 76-86.

Renjini, R., Gayathri, N., Nalini, A. & Srinivas Bharath, M. M. (2012). Oxidative damage in muscular dystrophy correlates with the severity of the pathology: role of glutathione metabolism. *Neurochem. Res., 37*, 885-898.

Rezvani, H. R., Rossignol, R., Ali, N., Benard, G., Tang, X., Yang, H. S., Jouary, T., de Verneuil, H., Taïeb, A., Kim, A. L. & Mazurier, F. (2011). XPC silencing in normal human keratinocytes triggers metabolic alterations through NOX-1 activation-mediated reactive oxygen species. *Biochim. Biophys. Acta, 1807*, 609-619.

Roat, E., Prada, N., Ferraresi, R., Giovenzana, C., Nasi, M., Troiano, L., Pinti, M., Nemes, E., Lugli, E., Biagioni, O., Mariotti, M., Ciacci, L., Consolo, U., Balli, F. & Cossarizza, A. (2007). Mitochondrial alterations and tendency to apoptosis in peripheral blood cells from children with Down syndrome. *FEBS Lett., 581*, 521-525.

Rodriguez, M. C., MacDonald, J. R., Mahoney, D. J., Parise, G., Beal, M. F. & Tarnopolsky, M. A. (2007). Beneficial effects of creatine, CoQ10, and lipoic acid in mitochondrial disorders. *Muscle Nerve, 35*, 235-242.

Scarlato, G. N., Bresolin, N. I., Moroni, I., Doriguzzi, C., Castelli, E., Comi, G., Angelini, C. & Carenzi, A. (1991). Multicenter trial with ubidecarenone: treatment of 44 patients with mitochondrial myopathies. *Rev. Neurol. (Paris), 147*, 542-548.

Scheibye-Knudsen, M., Ramamoorthy, M., Sykora, P., Maynard, S., Lin, P. C., Minor, R. K., Wilson, D. M. 3rd, Cooper, M., Spencer, R., de Cabo, R., Croteau, D. L. & Bohr, V. A. (2012). Cockayne syndrome group B protein prevents the accumulation of damaged mitochondria by promoting mitochondrial autophagy. *J. Exp. Med., 209*, 855-869.

Schöls, L., Zange, J., Abele, M., Schillings, M., Skipka, G., Kuntz-Hehner, S., van Beekvelt, M. C., Colier, W. N., Müller, K., Klockgether, T., Przuntek, H. & Vorgerd, M. (2005). L-carnitine and creatine in Friedreich's ataxia. A randomized, placebo-controlled crossover trial. *J. Neural Transm., 112*, 789-796.

Sparaco, M., Gaeta, L. M., Santorelli, F. M., Passarelli, C., Tozzi, G., Bertini, E., Simonati, A., Scaravilli, F., Taroni, F., Duyckaerts, C., Feleppa, M. & Piemonte, F. (2009). Friedreich's ataxia: oxidative stress and cytoskeletal abnormalities. *J. Neurol. Sci., 287*, 111-118.

Spurney, C. F., Rocha, C. T., Henricson, E., Florence, J., Mayhew, J., Gorni, K., Pasquali, L., Pestronk, A., Martin, G. R., Hu, F., Nie, L., Connolly, A. M., Escolar, D. M. & Cooperative International Neuromuscular Research Group Investigators. (2011). CINRG pilot trial of coenzyme Q10 in steroid-treated Duchenne muscular dystrophy. *Muscle Nerve, 44*, 174-178.

Sue, Y. M., Chou, H. C., Chang, C. C., Yang, N. J., Chou, Y. & Juan, S. H. (2014). L-carnitine protects against carboplatin-mediated renal injury: AMPK- and PPARα-dependent inactivation of NFAT3. *PLoS One, 9*, e104079.

Tanaka, J., Nagai, T. & Okada, S. (1998). Serum concentration of coenzyme Q in xeroderma pigmentosum. *Rinsho Shinkeigaku, 38*, 57-59.

Terrill, J. R., Radley-Crabb, H. G., Iwasaki, T., Lemckert, F. A., Arthur, P. G. & Grounds, M. D. (2013). Oxidative stress and pathology in muscular dystrophies: focus on protein thiol oxidation and dysferlinopathies. *FEBS J.*, *280*, 4149-4164.

Tiano, L. & Busciglio, J. (2011). Mitochondrial dysfunction and Down's syndrome: is there a role for coenzyme Q(10)? *Biofactors, 37*, 386-392.

Tiano, L., Padella, L., Santoro, L., Carnevali, P., Principi, F., Brugè, F., Gabrielli, O. & Littarru, G. P. (2012). Prolonged coenzyme Q10 treatment in Down syndrome patients: effect on DNA oxidation. *Neurobiol. Aging, 33*, 626.e1-8.

Torrioli, M. G., Vernacotola, S., Peruzzi, L., Tabolacci, E., Mila, M., Militerni, R., Musumeci, S., Ramos, F. J., Frontera, M., Sorge, G., Marzullo, E., Romeo, G., Vallee, L., Veneselli, E., Cocchi, E., Garbarino, E., Moscato, U., Chiurazzi, P., D'Iddio, S., Calvani, M. & Neri, G. (2008). A double-blind, parallel, multicenter comparison of L-acetylcarnitine with placebo on the attention deficit hyperactivity disorder in fragile X syndrome boys. *Am. J. Med. Genet. A, 146*, 803-812.

Valentin-Vega, Y. A., Maclean, K. H., Tait-Mulder, J., Milasta, S., Steeves, M., Dorsey, F. C, Cleveland, J. L., Green, D. R. & Kastan, M. B. (2012). Mitochondrial dysfunction in ataxia-telangiectasia. *Blood, 119*, 1490-1500.

Von Kobbe, C., May, A., Grandori, C. & Bohr, V. A. (2004). Werner syndrome cells escape hydrogen peroxide-induced cell proliferation arrest. *FASEB J., 18*, 1970-1972.

Wallace, D. C., Fan, W. & Procaccio, V. (2010). Mitochondrial energetics and therapeutics. *Annu. Rev. Pathol., 5*, 297-348.

Xia, X., Werner, D., Popanda, O. & Thielmann, H. W. (1994). Expression of mitochondrial genes and DNA-repair-related nuclear genes is altered in xeroderma pigmentosum fibroblasts. *J. Cancer Res. Clin. Oncol., 120*, 454-464.

Yen, H. C., Chen, F. Y., Chen, S. W., Huang, Y. H., Chen, Y. R. & Chen, C. W. (2011). Effect of mitochondrial dysfunction and oxidative stress on endogenous levels of coenzyme Q(10) in human cells. *J. Biochem. Mol. Toxicol., 25*, 280-289.

Zhang, Q. S, Marquez-Loza, L., Sheehan, A. M., Watanabe-Smith, K., Eaton, L., Benedetti, E., Major, A., Schubert, K., Deater, M., Joseph, E. & Grompe, M. (2014). Evaluation of resveratrol and N-acetylcysteine for cancer chemoprevention in a Fanconi anemia murine model. *Pediatr. Blood Cancer, 61*, 740-742.

Index

S

T